# The Final Frontier

# The Final Frontier

The Rise and Fall of the American Rocket State

DALE CARTER

VERSO

London · New York

First published by Verso 1988
© 1988 Dale Carter
All rights reserved

**Verso**
UK: 6 Meard Street, London W1V 3HR
USA: 29 West 35th Street, New York, NY 10001 2291

Verso is the imprint of New Left Books

**British Library Cataloguing in Publication Data**

Carter, Dale, *1957* —
    The final frontier, the rise and fall of
    the American rocket state.
    1. United States. Economic development.
    Related to space flight    2. United States.
    Space flight. Related to economic
    development.
    I. Title
    330.973'092

ISBN 0 86091 192 6 hbk
ISBN 0 86091 908 0 pbk

**US Library of Congress Cataloging in Publication Data**

Carter, Dale, 1957-
      The final frontier: the rise and fall of the American rocket
state/Dale Carter.
        p.      cm.
    Bibliography: p.
    Includes index.
    ISBN 0-86091-192-6.     ISBN 0-86091-908-0 (pbk.)
    1. Aetronautics — United States — History.    I. Title.
    629.13'00973 — dc19

Typeset by Leaper & Gard Ltd, Bristol, England
Printed in Great Britain by Biddles Ltd, Guildford

# Contents

# The Haymarket Series

Editors: Mike Davis and Michael Sprinker

The Haymarket Series is a new publishing initiative by Verso offering original studies of politics, history and culture focused on North America. The series presents innovative but representative views from across the American left on a wide range of topics of current and continuing interest to socialists in North America and throughout the world. A century after the first May Day, the American left remains in the shadow of those martyrs whom this series honours and commemorates. The studies in the Haymarket Series testify to the living legacy of activism and political commitment for which they gave up their lives.

*for Eric Mottram*


Is the cycle over now, and a new one ready to begin?
Will our new Edge, our new Deathkingdom, be the Moon?

Thomas Pynchon, *Gravity's Rainbow* (1973)


Space is the theater of strategy of tomorrow – space and
the human mind.

James Gavin, *War and Peace in the Space Age* (1958)


This operation is somewhat like the periscope of a submarine.

Michael Collins, Apollo XI Command Module
pilot (24 July 1969)

# Introduction

At the immediate risk of finding myself the most unpopular character of all fiction – and history is fiction – I must say this: 'Bring together state of news – Inquire onward from state to doer – '

— William Burroughs, *Nova Express* (1966)

In April 1964 President Lyndon Baines Johnson travelled from Washington to perform the official opening of what was at the time the largest global exhibition ever held, the New York World's Fair. Bringing together contributions from the federal government, dozens of the nation's leading private corporations, over thirty American states, and more than forty other countries, the fair was intended to promote and celebrate the fruits of corporate ingenuity and responsible public intervention, of freedom and democracy, as embodied in the diverse high technology products of the world's most advanced industrial society. The celebration retained a singularly futuristic air. As Walter McDougall emphasizes, its attractions included a 'Futurama' exhibition from General Motors, New York State's 'Tent of Tomorrow,' and a twenty-first century voyage under the guidance of the Westinghouse Corporation's 'Time Capsule.' More precisely still, it described one universally anticipated domain for all the potential civilizations these institutions portrayed. Whether in the Transportation and Travel Pavilion, which gave over its entire second floor to a simulated moonscape and a presentation of the Apollo lunar landing mission, in the Hall of Science, which featured Martin Marietta mock-ups demonstrating space rendezvous techniques soon to be perfected by the American space agency's latest Gemini program, or in the open air Space Park, which brought together the largest collection of rockets and spacecraft ever assembled outside Cape Kennedy, visitors to the exposition were left in little doubt of the shape of things to come. The future for the American pioneering spirit lay in the construction of new worlds on the final frontier of space. As if to underline this self-evident truth, the fair included among its most striking features a 140-foot high, hollow steel globe encircled by three symbolic orbits entitled the 'Unisphere.' What President Kennedy had two years earlier described as its primary theme, 'Man's Achievement in an Expanding Universe,' was thereby encapsulated in a 'homogeneous metallic world lassoed by technology.' Such was the promise of space travel.[1]

That American enterprise and democratic institutions would eventually give rise to some elevated, terminal, global society was, of itself, hardly a novel idea. Twenty-five years earlier, New York's previous World's Fair had included a scale model of a perfectly engineered 'World of Tomorrow' entitled 'Democracity' enclosed within another great globe dubbed the 'Perisphere.' Whereas the 1939 model was described by its promoters as a 'vast Utopian stage set,' the 200 pavilions and sideshows of the 1964 World's Fair together constituted but one installment in a much larger construction program that also included a massive renovation of Flushing Meadow, work on the Lincoln Center for the Performing Arts, and a number of other major civil engineering projects.[2] What had been celluloid fantasy less than thirty years before in *Things to Come* was by the mid-1960s being reproduced in steel and glass. When, within a year of the latter exposition opening, Walt Disney completed purchasing an area of Florida twice the size of Manhattan for the construction of an Experimental Prototype Community of Tomorrow (EPCOT), his

---

1. *New York Times*, 23 April 1964, pp. 1, 26-31; Walter A. McDougall, *The Heavens and the Earth* (New York: Basic Books, 1985), p. 399; *The Times* (London), 21 April 1964, p. 11, 22 April 1964, p. 14, 23 April 1964, p. 11. For details and photographs, see *New York Times Magazine*, 19 April 1964, Part II. This special 128-page supplement was given over entirely to the World's Fair. Entitled 'The Fair, The City, The Future,' it was divided into three parts and concluded with an essay by NASA Deputy Administrator Hugh L. Dryden on the future of the American space program beyond Apollo.

2. Mike Wallace, 'Mickey Mouse History: Portraying the Past at Disney World,' *Radical History Review* 32 (1985), p. 42; McDougall, p. 399. On the 1939 World's Fair, see Mel Scott, *American City Planning Since 1890* (Berkeley: University of California Press, 1971), pp. 360-63. On the full 1964-5 construction program, see New York World's Fair Corporation, *The Fair in 1965* (New York: NYWFC, 1965), pp. 31, 37-43.

decision only underlined the extent to which once 'utopian' stage sets were on the verge of coming to life. Disney's experimental city was to be 'a planned, controlled community, a showcase for American industry and research, schools, cultural and educational opportunities' whose integrated circuits would embrace 20,000 people within the protective walls of its own gigantic bubble dome. As Mike Wallace notes, until Disney's death in 1966 and the subsequent decision of his corporate reincarnation, WED Enterprises, to express their founder's last dream in a less idealistic form, the city was to make use of 'American knowhow, ingenuity, and enterprise' to substitute new ideas and practices for the more familiar ills of urban life, from slum housing to property speculation. It would, in effect, draw on the lessons of history and its material products to construct a concentrated social laboratory divorced from history, a city upon a hill, a man-made heaven on earth.[3]

In a formal sense at least, Disney's elaborate construction and its precursors would have provided ideal environments for Dominus Blicero, the German World War II SS officer portrayed in Thomas Pynchon's *Gravity's Rainbow* who, in common with many of the novel's characters, endeavors to locate himself within some universal plot or inviolable enclosure as a mechanism of personal security. Like these later engineers, Blicero 'dream[s] of a great glass sphere, hollow and high and far away' continuously evacuating itself and its inhabitants of the waste and destruction of history; like them, he seeks to mold future generations within his own terminal design; like them, he sees in the figure of the rocket a means of access to such a static, timeless, all-absorbing world. The analogy may appear incongruous but it is not. For the vital role played by Blicero in Pynchon's epic fiction of a world given over entirely to the politics of engineering lies precisely in his anticipating and facilitating the materialization a generation beyond of the post-war society apotheosized in these elevated worlds. When his desire to erect such a space age steady state

---

3. William Cameron Menzies (dir.), *Things to Come*, written by H.G. Wells, London Films, 1936; Wallace, pp. 41-2; Alexander Wilson, 'The Betrayal of the Future: Walt Disney's EPCOT Center, *Socialist Review* 15, no. 83 (1985), p. 41. On the extensive theory and practice of such construction in the 1960s, see Reyner Banham, *Megastructure: Urban Futures of the Recent Past* (London: Thames & Hudson, 1976); Paolo Soleri, *The City in the Image of Man* (Cambridge, Mass.: MIT Press, 1970); R. Buckminster Fuller and Robert Marks, *The Dymaxion World of Buckminster Fuller*, rev. edn (New York: Anchor Doubleday, 1973), pp. 231-5.

proves beyond his personal grasp, as it would prove for Disney, he too is replaced by a more long-lasting corporate agency – Pynchon's *Raketen-Stadt* – whose harbinger is a specially designed V–2 rocket inside which the rapidly fading Blicero stuffs his chosen replacement, a young rocket-mad *Wehrmacht* conscript named Gottfried. The launch of the rocket in the dying days of World War II is the means by which Blicero hopes to deliver his abiding dream across the post-war skies and into material form. It is a passover, as yet incomplete, which in a number of ways passes over – and reaches directly into – Disney's incipient world.[4]

That the space age future anticipated by both Dominus Blicero and the designers of the Experimental Prototype Community of Tomorrow had a more immediate terrestrial analog is evident enough. At EPCOT, as Wallace reports, '[p]assengers settle themselves into moving vehicles which carry them from the dim past to an imagined future.' But while that future 'is always set in outer space' and represented by 'awesome starry expanses in which space stations and satellites hover,' life in space 'looks remarkably like life on sitcom TV.'[5] Much the same could be said of Blicero's celestial future and of the previews of life inside the *Raketen-Stadt* distributed across the complex spaces of *Gravity's Rainbow*. Blicero himself tells Gottfried in his parting elegy before the launch of the final V–2 that the colonists he dreams of in the great glass sphere may 'have learned to do without air' but that 'they weren't really spacemen.' The daughter of the rocket engineer who spends hours dreaming of setting up home next to 'a small pretty crater in the Sea of Tranquillity'

4. Thomas Pynchon, *Gravity's Rainbow* (New York: Viking, 1973), p. 723.

5. Wallace, pp. 44-5.

6. Pynchon, pp. 410, 419-23, 723; Richard Schickel, *The Disney Version*, rev. edn (London: Michael Joseph, 1986), pp. 317-19. Pynchon's description of a holiday resort administered by children is probably based on the fact that after the original Disneyland opened in Anaheim, Orange County, Southern California, in July 1955, Disney instituted citizenship programs in which children acted as guides during holiday periods. It is possible that Pynchon, who was working for Boeing at Seattle, Washington, between 1960 and 1962, visited Disneyland and witnessed not only this particular scheme in action but also the popular moon rocket ride and monorail systems of 'Tomorrowland.' Similarly, the 1964 World's Fair in New York attracted massive publicity and Pynchon may either have visited the pavilions, witnessed the film shows, or read reports. In 1966, the year his previous novel, *The Crying of Lot 49*, was published, Pynchon was resident in New York City. See Schickel, pp. 315, 322-5; *New York Times Book Review*, 17 July 1966, pp. 22, 24.

has her imagination fired by annual visits to Zwölfkinder, Pynchon's prototype Disneyland, and will in practice find her ideal home amid the 'smiling lawns and tasteful cottages' of the post-war American suburb apotheosized at EPCOT. The efficient, friendly, and soothing attendant who reassures new arrivals in their movements across the burgeoning *Raketen-Stadt* towards the end of the novel could have come direct from EPCOT's own on-site university.[6] If Pynchon's text begins with an earlier generation of evacuees in basic training for life under the shadow of the rocket, pressed into less accommodating rail cars and moving off into the darkness of a nation at war, it prefigures their dispersal and relocation, not to lunar space stations but to the more proximate but equally hostile atmosphere of post-war American affluence; not to remote and imaginary glass spheres but to the skyscrapers, offices, and shops, schools and parks, sports coliseums and civic centers of high societies like Clear Lake City near the Johnson Space Center in Houston, and to other comparable material expressions of the real American space age which came fully into being from the late 1950s onwards.[7]

These terrestrial safe havens were, of course, no more divorced from history, no more inviolable, than their idealized prototypes. In *Nineteen Eighty-four*, Winston Smith had imagined the glass paperweight he and Julia acquired to be a tiny world complete with its own atmosphere in which 'time could be arrested,' a sanctuary 'belonging to an age quite different from the present one' under whose protective dome 'no harm could come to them.' It functioned not as a spell against Big Brother but as the instrument of his malicious design. Fifteen years later the opening of the New York World's Fair witnessed what *The Times* in London described as 'a good many well-organized protests' by civil rights groups denouncing segregation in schools and the construction industry as well as the state's decision to invest in white middle-class extravaganzas in preference to sorely needed public housing. Numerous violent confrontations, some 300 arrests, and a (thwarted) attempt to climb the much-publicized Unisphere ensued. In both cases the dream of inaugurating some terminal, invincible system into which one might retreat as if within a Spenglerian curve was shown to be not only futile but

---

7. On Clear Lake City, see Stephen B. Oates, 'NASA's Manned Spacecraft Center at Houston, Texas,' *Southwestern Historical Quarterly* 67 (1963/64), pp. 374-5.

sacrificial. By the time EPCOT finally opened as part of the Florida Disney World in 1982, it had been converted from a utopian community into a sound business proposition: an imaginative combination of patriotic diorama, popular amusement park, industrial exposition, and corporate celebration designed by General Motors, Kodak, and their peers to captivate the American public as loyal and satisfied customers without having to underwrite their liberation as informed and intelligent citizens. But reformulating Walt Disney's original escape from history as an introduction to a series of sophisticated corporate versions of history, which maintained assurance only by excluding facts, meant more of the same. Two years later, when WED Enterprises found itself subject to the designs of Wall Street financiers and Texan corporate raiders, its efficient, clean, and friendly labor force were obliged to carry the can through real wage cuts and enforced redundancies. As Blicero reminds Gottfried before his own compulsory evacuation in *Gravity's Rainbow*: 'Gravity rules all the way out to the cold sphere, *there is always the danger of falling*.'[8]

This book is an investigation of these intersecting processes of evacuation, elevation, and elimination, of what Thomas Pynchon describes as the 'great frontierless streaming' which slowly decants the displaced representatives and products of a 'European and bourgeois order they don't yet know is destroyed forever' into a post-imperial holding tank being established across the Atlantic.[9] Moving from the ruined and rifled wastelands of Nazi Germany in 1945 to the elaborately constructed projects of the American space agency a generation and more beyond, it endeavors to detail one of the central propositions of *Gravity's Rainbow*: the replacement of a wartime 'Oven State' dominated by Dominus Blicero with a post-war 'Rocket State' under the long shadow of the Apollo XI spacecraft that in 1969 would carry the first men to land on the moon.[10] Rather than taking the evacuation with which *Gravity's Rainbow* begins as its point of

---

8. George Orwell, *Nineteen Eighty-four* (1949; Harmondsworth: Penguin, 1954), pp. 80, 120, 124; *New York Times*, 23 April 1964, pp. 1, 28; Wallace, pp. 39-43, 54-7; Pynchon, p. 723. Pynchon's allusion is also of course to the Puritan fall, perhaps most celebratedly defined in Jonathan Edwards's 'Sinners in the Hands of an Angry God.' See Jonathan Edwards, *Representative Selections*, ed. Clarence Faust and Thomas H. Johnson (New York: Hill & Wang, 1962), pp. 155-6.

9. Pynchon, pp. 549, 551.

departure, however, *The Final Frontier* enters the field of the novel at its literal, structural, and historical center: the 'Terminal' conference at Potsdam where in July 1945 Truman, Churchill, and Stalin met to tie up the loose ends of war and to sketch out the pattern of peace. For it is around the focal point of Potsdam that perhaps the pivotal character of this 'radically decentered' novel, Tyrone Slothrop, undergoes the complex set of transformations whose completion indicates the supersession of the power structure associated with Nazism. Correspondingly, it is as he directs Slothrop's movements across the seemingly chaotic terrain of *Gravity's Rainbow* and beyond the great divide of Potsdam that Pynchon describes the lineaments of the power structure associated, as Daniel Bell might have put it, with no 'ism': the incipient high society of Henry Luce's 'American Century.'[11]

This book is not, however, a piece of detailed literary criticism in the conventional sense. Rather, it draws on a variety of works on recent history, and in particular on the US space program, to rehearse some of the main issues of post-war American society. Thus, whereas the work begins with an excursion through the pages of *Gravity's Rainbow* in search of the Rocket State's material and imaginative roots, it then travels across a great deal of extra-literary space in order to locate and describe that state's historical culmination and terminus. The plot moves from Norman Mailer's *Of A Fire on the Moon* to Jacques Ellul's *Propaganda*, from Mary Holman's *Political Economy of the Space Program* to Robert Coover's *The Public Burning*, from Edward Thompson's *Star Wars* to Nora Sayre's *Running Time: Films of the Cold War*. Nor, it must be emphasized, is this an exercise in interdisciplinary studies. Rather, the procedure is one of synthesis, its necessity being to document events in dynamic interaction and to reveal the pattern of interaction in the culture, not to establish some hierarchical order of influence or singular perspective. It could hardly be anything else. The rocket which moves from Nazi Germany to the United States, from the *Wehrmacht* to the American space

---

10. *Raketen-Stadt* literally means 'Rocket City,' but since Pynchon also refers implicitly and explicitly to the formation of a 'Rocket State' the latter designation will be adopted. In the elaborate architecture of Pynchon's nascent world, the former is the dynamic pole and archetypal instance of the latter.

11. Paul Coates, 'Unfinished Business: Thomas Pynchon and the Quest for Revolution,' *New Left Review* 160 (1986), p. 124; Daniel Bell, *The End of Ideology: On the Exhaustion of Political Ideas in the Fifties* (New York: Collier Books, 1961).

agency, from the UFA film studios at Neubabelsburg to the 1964 World's Fair and Disneyland, certainly provides a focus for the materials drawn into the composition, but these materials have no fixed boundaries, either of artistic form or of subject matter, and the focus is, by definition, mobile. The intention is, therefore, to facilitate the construction of a set of informational intersections jointly capable of handling the complex grip of the space program within post-war American society.[12]

The work falls into three distinguishable parts, each comprising two chapters. The first investigates the formation and growth of the social order perhaps most fully realized – notwithstanding the region's simultaneous fostering of the John Birch society – in the official Southern Californian lifestyle of the 1950s: Pynchon's perpetual 'Happyville.'[13] Chapter 1 makes use of a handful of figures from *Gravity's Rainbow* to dramatize the movement from 'Oven State' to 'Rocket State' as a process of absorption facilitating the survival, transformation and reproduction of a partly obsolete imperial power structure in the form of its incipient totalitarian replacement. Drawing on the works of Hannah Arendt, Wilhelm Reich, and more recent writers, it rehearses the struggle between Allied and Axis forces during World War II as a conflagration in which two related but distinct control structures – one adaptive and victorious, the other overadapted and vanquished – forge a novel system of control characterized at its civilized center by an abandonment of military methods in favor of more efficient, less bloody techniques. Through Pynchon's fictional characters it anticipates the production of that system's associated society, while through the historical figures of German industrialist Walter Rathenau, Soviet diplomat George Chicherin, and American President Franklin D. Roosevelt it follows the development of its preferred elite and of the hegemonic 'Operation' their activities advance.

Chapter 2 then investigates the new forms of conflict and control characteristic of the ostensibly post-imperial era. By noting some of the historical contexts of incipient totalitarian society's development in the United States between 1945 and 1957, it dramatizes that society as a precarious structure of

---

12. On this issue see Eric Mottram's important essay, 'Developing Cultural Studies and American Studies,' *Journal of Area Studies* 14 (1986), pp. 6-12.

13. Mike Davis, *Prisoners of the American Dream* (London: Verso, 1986), pp. 158-60: 195; Reyner Banham, *Los Angeles* (1971; Harmondsworth, Penguin, 1973).

immanent civil conflict and insatiable security programming, of evasive action and endemic stasis: a high society installed in a no man's land within which a number of other figures liberated from the ruins of Pynchon's Oven State – notably technical experts, businessmen, and propagandists – extend their operational grip. The liberal social, political, and economic mechanisms these figures adopt are presented as the means by which the outdated and untenable relationships of authority and obedience characteristic of the imperial age are updated and reproduced in the interests of the post-war order's extension. Where orderly victory constituted the objective of the now vanquished Oven State, its descendant's ambition is characterized as permissive security, while the endlessly absorbing expression of such a movement is presented as the rocket, translated from its material expression under the Nazis – the V–2 – into its more imaginative forms at the heart of the Rocket State's rise.

The second part of this study investigates the maturation and apotheosis of the Rocket State through the careers of its prime exponents. Chapter 3 illustrates how, by penetrating that state's still precarious foundations between 1957 and 1961, the orbits of the Soviet Sputniks and of Yuri Gagarin paved the way for its subsequent integration: boosting the career of John F. Kennedy, the political embodiment of incipient totalitarianism, and making a manned lunar landing program one of the latter's most significant proposals. Through a consideration of his personal marketing, his electoral strategy, and his political necessities, Kennedy is defined as part of the Operation's multifaceted replacement for its outdated *Führer* system: a figure capable of addressing an appeal for voluntary totalitarianism across an indulgent society by calling for a more accommodating joint command structure.

Chapter 4 goes on to analyze the visible expressions of this structure: the astronauts who were intended to give the US space program – and the American President – an appeal extending across the whole of the political spectrum. Beginning with the first seven Mercury trainees and ending with the three men chosen to execute the initial lunar landing, it considers them as complex representatives of a society defined by its endemic conflicts: as component parts within a well-integrated operational structure, charismatic stars within a culture responsive to Hollywood codes, and sacrificial lambs within a society attuned to disposable lines. It then proceeds to rehearse the launch of Apollo XI as the ultimate expression of both the Rocket State's joint command structure and of its uniformly seductive appeal: a celebration of liberty and security, of power and surrender, of escape and refuge produced through the combined efforts of those technical experts, businessmen, and propagandists called to the Rocket State's aid. Finally, it contrasts the chimerical nature of this public celebration of American life with the private gains fed back to the Operation's hired hands, using the career of John Glenn between capsule and Capitol Hill as an instance of the power elite's reproduction.

The third and final part locates the preceding exposition within a deeper set of material contexts that first substantiate, then qualify, and finally explode the Rocket State's claims. The first half of chapter 5 illuminates the industrial, social, and economic moments of the post-war order's hegemony by considering the community of interest mobilized around the Mercury, Gemini, and Apollo programs during the 1960s: the role of the aerospace industry in supplying the hardware of space, of a diverse population of subcontractors in servicing the needs of these giants, and of a workforce hundreds of thousands strong in building and assembling the millions of parts that together took men to the moon. It then goes on to detail the space program's extension of the Rocket State's realm into previously unreconstructed regions – in this case the southern United States – before enumerating the terms of its most far-reaching claims: the diverse range of benefits space exploration promised to bring to mankind as a whole. The second half of the chapter starts to undermine these apparent strengths by exposing their qualified grounds. It redefines the pan-classical alliance created by the American space program as a strictly temporary concert of interests brought about as a result of a post-war capitalist growth phase whose limits were already evident by the launch of Apollo XI in July 1969. It contrasts the state's celebration of the responsible private enterprise system ostensibly fostering national space achievements with the business community's actual exploitation of state resources in pursuit of monopoly gain. And it emphasizes the paucity of the moon landing's benefits for the world in whose name it was planned.

The conclusion then equates the end of the Apollo program with the accelerating fragmentation of the Rocket State itself in the late 1960s and early 1970s. Rather than considering the first manned space program as simply a disposable product of the 'mad competition for status' that characterized a now abandoned phase of the Cold War, however, it sketches out the relationships between the development of NASA's second, shuttle-dominated space program and the onset of the second Cold War.[14] By investigating the expansion of secret military space operations carried out by the US government behind the screens of the Rocket State's more celebrated civilian efforts, it delineates the reassertion of that armored imperative previously absorbed within incipient totalitarian society's casual folds. By unveiling the Operation's response to the contradictions undermining the Rocket State's existence it defines the elevation of Ronald Reagan and of his prospective 'Star Wars society' as the power elite's solution to the disarticulation of the post-war order: a solution whose mad dream of libertarian refuge may secure no more than the end of the world.

Work on this study began with the assistance of grants from the Department of Education and Science for which I am most grateful. During the course of writing I have benefited in a variety of ways from the assistance of a number of people and other institutions. The University of Warwick has provided both employment in the Joint School of Comparative American Studies and an audience for parts of what follows. Senate House and the British Library in London furnished many of the sources on which I have drawn – in particular their collections of back issues of the *New York Times* – while the British Library Lending Division at Boston Spa gave me access to a number of otherwise unavailable texts. The relevant offices of Boeing Aerospace, North American Rockwell, Lockheed, McDonnell Douglas, IBM, and Grumman, in the United States, all responded to my enquiries promptly, as did William Johnson and Dan H. Fenn, Jr. at the John F. Kennedy Library in Boston, Martin Elzy and Claudia Anderson at the Lyndon B. Johnson Library in Austin, and the staffs at NASA's Scientific and Technical Information Facility at Baltimore and the Johnson Space Center in Houston. I am particularly grateful to Lois Lovisolo of the History Center at Grumman Aerospace for materials on the construction of the Apollo Lunar Module. In England, David Smith supplied me with technical analyses of the Strategic Defense Initiative, and Joey Smith loaned out his collection of Pynchon criticism. Clive Bush of the Department of English, University of Warwick, and Professor Peter Parish, Director of the Institute of United States Studies, University of London, both read the original manuscript and

---

14. See *New York Times*, 25 May 1972, p. 44.

made a number of suggestions for its improvement. Peter Burford generously advanced a sizeable loan without which the work could not have been completed.

I have also received much assistance in preparing the text for publication. Neil Belton at Verso took the first risk, reading drafts of early chapters and encouraging me to produce more. Thereafter, Mike Sprinker, General Editor of the Haymarket series in New York, provided a detailed and sympathetic commentary on the text as well as recommendations for its expansion. Mike Davis in London shared his unrivalled knowledge of the political economy of the Californian aerospace industry with me and, through his own contribution to the Haymarket series, informed my understanding of the Rocket State's disarticulation. Beyond Verso's editorial circuits, John King of the Joint School of Comparative American Studies, University of Warwick, has from the outset been instrumental in advancing the production of this book.

I remain most indebted, however, to a pair of long-term teachers. Callum MacDonald rekindled my original interest in the US-manned space program and has for over a decade now provided hospitality, encouragement, and advice. Eric Mottram furnished, through both his writing and teaching, the comprehensive grounding in cultural studies without which this essay could not have been written, the guidance which encouraged me to transform an interest in Thomas Pynchon into a more or less ordered exposition, and an exemplary critical intelligence that has sustained my own efforts from beginning to end. I hope the following discharges some part of my accumulated debts. Finally, I should like to thank Ala Szczepura for driving me in.

Dale Carter

Leamington Spa
August 1987

# 1

# Over the Rainbow

After the act of wounding, breaking, what's to become of the little Oven-state? Can't it be fixed? Perhaps a new form, one more appropriate ... the War itself as tyrant king ... it can still be salvaged can't it, patched up, roles reassigned, no need to rush outside where ...

— Thomas Pynchon, *Gravity's Rainbow* (1973)

Man's ... entire modern history has been characterized by a series of breakdowns and blurrings of boundaries. And since World War II that process has become so intense and so extreme that we may take the year 1945 as the beginning of a new historical epoch.

— Robert Jay Lifton, *Boundaries* (1970)

In a series of close-ups and panning shots from Thomas Pynchon's *Gravity's Rainbow*, American lieutenant Tyrone Slothrop zigzags, skulks, and wades his way across the autobahns and lakes of occupied Berlin. Already masquerading as English war correspondent Ian Scuffling, Slothrop is on a covert and potentially lucrative mission to recover six kilos of top quality Nepalese hashish, stolen by 'semi-AWOL' American Seaman Bodine and temporarily buried beneath an ornamental juniper bush. The bush is located right in the center of what is presently the compound housing Allied delegates to the Potsdam conference. Security is tight – as Bodine puts it, 'you couldn't get a gnat in there now' – and Slothrop can only bluff his way past the guards by adding new elements to his established dual personality. Just as one narcotic product conceals another, so Slothrop accumulates a deepening palimpsest of aliases and disguises. A forged identity card, trip ticket, and conference passes come from

Gerhardt von Göll, celebrated black marketeer and film director, giving Slothrop another name, Max Schlepzig, which also happens to be the name of one of von Göll's pre-war film stars. A disguise of sorts comes from Emil Bummer, former cat burglar and middleman in the hashish deal, who provides a velvet cape, buckskin trousers, and a pointed helmet stolen from a batch of Wagnerian opera costumes. Bummer dubs Slothrop 'Rocketman,' adding to his Wagnerian image the role of comic-strip caped crusader, a second cousin to Slothrop's beloved Plasticman, and suitable persona for his mission impossible to recover the hashish. In the circumstances Slothrop appropriately 'decides to pose as a vaudeville entertainer, an illusionist.' It is something he has experience of.[1] These disguises are carefully adopted, not so much by Tyrone Slothrop as by Thomas Pynchon. In *Gravity's Rainbow* he presents Slothrop as a picaresque hero, the direct descendant of the protagonist in Voltaire's *Candide* (1759) and of the Cosmopolitan in Herman Melville's *The Confidence Man* (1857). And he does so for familiar reasons. As he steers Slothrop towards his goal, Pynchon uses this precisely engineered multiple identity as the turning point of his satire on the relationships among disguise, fragmentation, and power inside incipient totalitarian society.[2]

Slothrop's destination is as stratified and deceptive as his appearance. The Allied delegates' compound is in Neubabelsburg, just across the Havel canal from Potsdam itself. The very name of the suburb, which means 'new speech town,' suggests a transformation in the structure of language, communication, and therefore control, an issue dramatized by George Orwell at the time in *Nineteen Eighty-four* (1949). Tyrone's objective thus proves to be a compound in more ways than one. For while the delegates meet to thrash out the post-war European order, German reparations, and Japanese surrender terms, these political and military issues seem overshadowed by the theatrical glamor of the Allied housing center. 'The whole joint is lit up like a Hollywood premiere,' is Slothrop's initial reaction: 'They must deal here with a strange collection of those showbiz types.'[3] The

---

1. Thomas Pynchon, *Gravity's Rainbow* (New York: Viking, 1973), pp. 365-72, 377-80. Subsequent references, all to this edition, will be placed within the body of the text. To avoid unnecessary repetition citations will be grouped together and only major quotations noted individually.
2. On the terms 'totalitarian' and 'incipient totalitarian', see Appendix 1.

imagery, rooted in the fact that before the war Neubabelsburg had been 'the old movie capital of Germany,' is hardly coincidental. As C. Wright Mills notes, the power of the 'showbiz type' image, which by the start of World War II had largely replaced the figure of the café society debutante as the focus of public attention, derived from the unrivalled influence of Hollywood in the 1930s and 40s.[4] Slothrop's response is thus a representative one, drawing on the dominant public image bank of the time. As such it constitutes a useful first intersection point within Pynchon's complex drama (*GR* 371, 380).

At the Allied housing compound Slothrop is propelled into what Mills terms 'a kaleidoscope of highly distracting images.' He moves within a propagandized environment.[5] To decode these images, and thus to illuminate the structure and meaning of the deception, Pynchon uses the equally deceptive Slothrop in two related ways: as the agent of satire and as its naive victim. Tyrone's initial reactions neatly exemplify his function. His Hollywood-related phrases, whilst ostensibly inaccurate, are also the *lapsus linguae* that expose other truths. Thus he may not be witnessing a film premiere, but if the Allied housing compound is, as Tyrone says, a 'joint,' then it becomes at once a center of illicit public entertainment, a form of narcotic, and an intersection point between related structures. It is revealed, in other words, as a significant multiple image. Again, calling Truman, Churchill, and Stalin 'a strange collection of those showbiz types' may be an apparent error, but it can also be taken as a mordant and neatly accurate caricature. Slothrop's choice of words and images thus gives access to the main issues in the very process of his deception: continuing a long literary tradition, he is the fool who speaks the truth.[6]

As Tyrone moves across the compound, Pynchon deepens the field of the image. The hashish is buried outside a villa whose 'stenciled alias' reads THE WHITE HOUSE. Inside, Slothrop can hear a dance band, singers, laughter, and 'multilingual chitchat.'[7] It is 'your average weekly night here at the great Conference.' This at least provides good cover for the caped illusionist.

3. Charles Clerc, 'Film in *Gravity's Rainbow*,' in Charles Clerc (ed.), *Approaches to Gravity's Rainbow* (Columbus: Ohio State University Press, 1983), pp. 129-33.

4. C. Wright Mills, *The Power Elite* (1956; New York: Oxford University Press, 1959), pp. 79-80.

5. Mills, p. 92.

6. Norman O. Brown, *Love's Body* (New York: Vintage, 1966), p. 244.

Through the french windows Slothrop sees someone who 'looks a bit like Churchill' and perhaps gets a glimpse of President Truman ('that guy in the glasses'), but he is unsure. The only 'showbiz type' Slothrop instantly recognizes is Mickey Rooney, who comes out onto the White House terrace for some air as Slothrop unearths the hashish. Tyrone is tempted to accept these images at face value. If true, they would explain his ease of entry – 'Are they expecting this magician, this late guest?' – and might also offer the prospect of a secure future: 'Glamour, fame. He could run in and throw himself at somebody's feet, beg for amnesty. End up getting a contract for the rest of his life with a radio network, o-or even a movie studio!' But they are false and Tyrone's reverie is sharply interrupted when he is confronted by barbed wire. The place is 'alive with roving sentries,' not talent scouts, so that Slothrop is reduced to cowering 'up to his ass' in a lake to avoid them (*GR* 381-2).

Tyrone's weakness for comic strips and the movies is thus the apparent target of this satire. By engaging the historical through the mediation of deceptively familiar mythology inside personal inexperience, he suffers the consequences. Faced with armed guards, his dreams of stardom are literally and unceremoniously swamped in the scramble for cover. Yet the final effect of this scene is rather different, for by lampooning Slothrop's susceptibilities Pynchon necessarily indicates the narcotic nature of power disguised as permitted glamor. The Potsdam conference is not a Hollywood premiere, but by reading it this way Slothrop illustrates the function of the imagery: once politics is taken as show business, the exercise of power can be translated into innocent, ahistorical news; politicians may become celebrities, their calculated rant magical charms justified by natural authority. Thus whilst Tyrone smokes this 'lit up' Neubabelsburg joint, in the process he smokes it out. Moreover, Pynchon's analysis does not satirize Slothrop alone. On the contrary, it next centers on the Allied housing compound guards. For while Tyrone is temporarily seduced by Neubabelsburg's appearance, the attraction is

---

7. Both Potsdam and Neubabelsburg escaped major damage during the war. The Allied delegates' housing had previously belonged to the UFA film organization whose employees had been, in Sir Alexander Cadogan's words, 'turned out.' Truman gave a reception for Stalin, Churchill, and Attlee on the evening of 19 July; US Secretary of State James Byrnes entertained with some Chopin piano music. See David Dilks (ed.), *The Diaries of Sir Alexander Cadogan, O.M., 1938-45* (London: Cassell, 1971), pp. 761-2, 767.

countered not only by the barbed wire but also by the job in hand. Hollywood dreams may feed his Rocketman fantasy but they cannot feed him: he still has to recover the hashish.

Here Pynchon uses Slothrop's own disguises to make him no longer a victim but an agent of the satire. Tyrone's access to the housing compound is dependent on the same structure of illusion that leads him to perceive power as neutral stardom: he outwits the security guards by hiding his military origins inside disguises and forged documents. Just as Truman and Churchill appear as posters and Hollywood partygoers, so Slothrop plays the magician. He becomes a showbiz type amidst showbiz types. But the composition of multiterraced Tyrone is designed to illuminate the changing structure of power and its associated images, not, as is the case with Truman and his friends, to deny it altogether. Each new layer in his engineered identity further mediates Slothrop's past: from American Army lieutenant Tyrone Slothrop he becomes, in succession, English 'war (peace?) correspondent' Ian Scuffling, German film star Max Schlepzig, and cartoon character Rocketman (*GR* 283, 366, 377). In these accumulating reflections Slothrop's military background moves inside, his nationality shifts and dissolves, and his work becomes reportage, act, and comic-strip fantasy. At the same time the structure of each image changes in line with the historical development of media technology: first, direct experience, followed by the printed word, moving representations of selected dramatized events, and finally visual images independent of any historical or natural basis. This multiple identity is used by Pynchon as the axis of his satire in the same way that Melville uses the Cosmopolitan in *The Confidence Man*. Slothrop is not so much an imposter himself as an agent of exposure: his masks are used to unmask others. Through Tyrone's layered appearance, Pynchon illuminates his environment, deploying him, in Melville's words, as though he were 'a revolving Drummond light, raying away from itself all around it.' By incorporating in Slothrop successive forms and images of authority, converting an historical process into a human figure, Pynchon can investigate the mythical behavior they engender.[8]

Slothrop is able to outwit the guards around the Allied housing compound because they are still governed by images

---

8. Herman Melville, *The Confidence Man: His Masquerade* (1857; New York: Signet, 1964), p. 246.

from the past. While he appears as 'the invisible youth, the armored changeling,' the guards remain addicted to maps of consciousness whose irrelevance holds them powerless:

> *Their* preoccupation is with forms of danger the War has taught them – phantoms they may be doomed now, some of them, to carry for the rest of their lives. Fine for Slothrop, though – it's a set of threats he doesn't belong to. They are still back in geographical space, drawing deadlines and authorizing personnel, and the only beings who can violate their space are safely caught and paralyzed in comic books. They think. They don't know about Rocketman here. They keep passing him and he remains alone, blotted to evening by velvet and buckskin – if they do see him his image is shunted immediately out to the boondocks of the brain where it remains in exile with other critters of the night. (*GR* 379)

The guards are, as ever, fully prepared for the last war. While they remain alert for enemy soldiers, patrolling space, time, and identity according to the standing orders of imperial power, Tyrone slips through their grid. His mobility across the 'joint' illuminates the guards' rigidity inside outdated structures. The use of checkpoints and armies is revealed as the strategy of a Canute, since Slothrop moves according to field actions, the electrical impedance of anxiety intersecting the magnetic attraction of stardom and hashish to determine his motion. When the sentries move to restrict lines of communication to the linear form and to monopolize them – their boxes are piled full of confiscated cameras while 'mimeographed handouts clog the stones and gutters' – their programming is exposed by Tyrone's mobility and the guards' reactions. At a time when science fiction comic-book and film-strip figures like Superman and Batman are flooding the image bank, the sentries read Slothrop accordingly. He is seen as simple entertainment, not an enemy soldier, and therefore not a threat (*GR* 379-80).[9]

Consequently, Tyrone reaches the hashish undetected by the guards. But his success is only one indication of the more fundamental change accompanying the end of World War II which is investigated in the transactions among Slothrop, sentries, and

---

9. The sentries are, of course, Russians. But Soviet comics of the 1930s and 40s included superheroes identical to the American breed. See John Griffiths, *Three Tomorrows: American, British, and Soviet Science Fiction* (London: Macmillan, 1980), p. 170.

compound. At the Allied housing compound in Neubabelsburg, Pynchon dramatizes events surrounding the cessation of hostilities and the Potsdam conference as indices of a crucial transformation in the nature and appearance of power: in the changing relationships between representation and perception, war and peace, history and myth, he presents the restructuring of the social, psychological, and visual forms of power which reveal both the passing of one control system – focussed in the guards – and the inauguration of another. The method is one of satirical exposure. In the same way that Herman Melville ruptures one hypothesis with another in *The Confidence Man*, at Neubabelsburg Pynchon subverts and decodes all the established orders of representation which are thus revealed as the coordinates of authority in transition: the guards carry guns but are powerless to stop Slothrop; the compound houses politicians but looks like a film set; Tyrone is an American Army lieutenant yet passes as a German civilian; Potsdam marks the end of hostilities and the victory of freedom yet is crawling with soldiers and swept by searchlights. We are at the junction between two related but distinct structures of power which this joint also embodies: in a formal, historical sense, between World War II and the Cold War which emerges from its ashes; in a deeper sense, between what Hannah Arendt terms imperialism and totalitarianism.

## When Paranoid Meets Paranoid

> Benway's first act was to abolish concentration camps, mass arrest and, except under certain limited and special circumstances, the use of torture.
>
> — William Burroughs, *The Naked Lunch* (1959)

In *The Origins of Totalitarianism*, first published in 1951, Arendt gives a detailed analysis of the foundations, development, and operation of imperial and totalitarian systems of power. In the process she identifies the fundamental changes which characterize the emergence of totalitarianism in the twentieth century as a control structure related to but distinct from imperialism. Although her discussion centers on Nazi Germany and Stalinist Russia, it retains equal relevance to those societies which dominate the 'post-totalitarian' world.

Summarizing these changes, Arendt recalls four characteristics of totalitarianism which emerged between 1933 and 1953:

No matter what the specifically national tradition or the particular spiritual source of its ideology, totalitarian government always transformed classes into masses, supplanted the party system, not by one-party dictatorships, but by a mass movement, shifted the center of power from the army to the police, and established a foreign policy openly directed toward world domination.[10]

The totalitarian struggle against bourgeois society which generated these transformations consisted of two distinct phases. In their rise to power totalitarian movements relied on a combination of propaganda and organization. Once installed they instituted and maintained a system of total domination through the use of terror in the service of an ideology. Necessarily, totalitarian governments transformed their domains into separate, sealed zones of strict ideological consistency, circumscribed by violence and extermination (blitzkrieg and the death camps) and regimented no longer by overt propaganda and physical coercion but through an automatic obedience resulting from the interiorization of command. Arendt argues that their emergence and eventual triumph entailed transformations not simply in the bourgeois social and political order but in its very psychology, what she calls 'the transformation of human nature itself.' It entailed, more precisely, a restructuring of consciousness as both constituting and constituted in relation to its social and political field. Thus totalitarianism in power, in its drive towards global domination, witnessed the decline of the nation-state, the extension of statelessness, and the associated loss of rights of residence and government protection; while in its related erosion of the class and party system it engendered the dissolution of established bourgeois individualism, the destruction of privacy and critical sensibility, the replacement of public involvement by indifferent passivity and loneliness, and the creation of a mass of indoctrinated units 'for whom the distinction between fact and fiction ... and the distinction between true and false ... no longer exist[ed].'[11]

Arendt's analysis provides one of many historical contexts for *Gravity's Rainbow*, since in his movement towards Neubabelsburg and across the joint Tyrone Slothrop both embodies and illuminates the dynamics of the transformation she investigates. In

---

10. Hannah Arendt, *The Origins of Totalitarianism*, new edn (New York: Harcourt Brace Jovanovich, 1973), p. 460.

11. Arendt, pp. xxxiii, 341-479 passim.

relation to the Allied housing compound Slothrop records the dissolution of the imperial order and at the same time suffers and incorporates the emerging totalitarian order. But whereas Arendt's original study takes Stalin's and Hitler's regimes as totalitarian termini, Pynchon implicitly dramatizes World War II as a period of accelerated transition, and focusses on the Potsdam conference as a turning-point in the maturation and long-term stabilization of an incipient totalitarianism which emerges from the conflict between the Axis and the Allies ending in 1945 and which inherits and synthesizes characteristics from all sides in a new hybrid power structure.[12] Therefore, while important, Tyrone's recovery of the hashish is not in itself a conclusion: it records the limits of imperialism but gives only partial access to its successor. The full transfer of power is only routinized, to borrow a term Pynchon adopts from Max Weber, with Slothrop's exit from Neubabelsberg, an exit that encapsulates the inauguration of the incipient totalitarian era.[13]

The dissolution of the imperial system is foreshadowed in Tyrone's approach to Neubabelsberg. As he heads from Nice through Switzerland and on into the Zone of occupied Germany, with the war in its final weeks and troops redeploying to the Far East, Slothrop senses the inherited structures of power dissolving. Approaching Zurich he feels that there is 'never a clear sense of nationality anywhere, nor even of belligerent sides.' The Greater Germany which had once engulfed Europe and realized in part the dream of global domination has clearly left its mark. What remains is a collection of 'rather stuffy conventions' like 'neutral Switzerland,' 'liberated France,' or 'totalitarian Germany,' observed but practically meaningless. The nation-state as a central unit of power is, at least in Europe, being bypassed (*GR* 257).[14]

At the same time, Tyrone himself experiences a transformation, at once political, social, and psychological, which embodies the fragmentation of the imperial order and hence the dispersal of its energies. Having decided to go AWOL from his

12. See the discussion of this point in Appendix 1.

13. On Weber and routinization, see Raymond Aron, *Main Currents in Sociological Thought 2, Durkheim, Pareto, Weber* (1968; Harmondsworth: Penguin, 1970), p. 245.

14. Cf. the remarks of George Kennan cited in Daniel Yergin, *Shattered Peace: The Origins of the Cold War and the National Security State* (1977; Harmondsworth: Penguin, 1980), p. 473, n. 13.

job with the Political Warfare Executive (PWE) at the Casino Hermann Goering on the French Riviera, Tyrone effectively slips the imperial network. Once he leaves the casino in April 1945 he becomes stateless, a condition underlined when he denies possessing either a 'people' or a 'country,' and loses his right to government protection and residence.[15] Indeed, his official World War II 'allies' move to deny him those rights. En route to his eventual destination Slothrop makes use of a succession of stolen or borrowed trappings to avoid recapture: a white zoot suit and identity card from AWOL forger Blodgett Waxwing; some workmen's clothes from Waxwing's agent in Zurich, a Russian named Semyavin; a pair of boots belonging to Soviet intelligence officer Vaslav Tchitcherine; an 'English English' accent from Foreign Office agent Sir Stephen Dodson-Truck and (strangely) Dutch triple-agent Katje Borgesius; and a moustache, which he thinks is his own idea. In the process he rehearses the dissolution of the power structure characteristic of the imperial order. Each new disguise he adopts represents a distinct imperial nation, class, race, or party, from rebellious young American blacks to fading English aristocrats; as he enters the Zone he effectively processes them and is in turn processed into a new hybrid identity whose nature accommodates and condenses their various traits.[16] He performs a function similar to that of the Cosmopolitan in Melville's *The Confidence Man*, adopting, presenting, and amalgamating facets of the structure of power. This new persona displaces the bourgeois identity he retained at the casino. There, as an Army lieutenant and bureaucrat, Slothrop epitomized the imperial character with its duties, rights, and procedures guaranteed by nationality and role. Now, stateless and without rights, his bourgeois identity is broken down in order to be replaced by its mass descendant. Where selfhood had once been guaranteed by nationality, at Neubabelsburg Tyrone requires an identity card to cross the joint: his very existence is a function of and depends on the permission of an alien authority imposing alienated consciousness. As he accumulates new identities Slothrop abandons his role within the military bureaucracy to become, as Ian Scuffling, a passive viewer of the last great classical

---

15. Slothrop's dream of begging Potsdam delegates for amnesty is useless, therefore. The very rise of the stateless leads to the abolition of the right of asylum. See Arendt, pp. 280, 293-4.

16. Cf. Arendt, p. 314.

imperial power's decline, and then, as Max Schlepzig and Rocketman, an uncritical toy or tool for the entertainment of the new authorities meeting at Potsdam. He becomes, that is, a partly willing and partly unwitting plaything of power. (*GR* 210-92, 363, 376)[17]

But Slothrop's liberation from the imperial order is only one moment in the restructuring of power. Its completion coincides with his integration into the new order which itself matures only as the novelty of Neubabelsburg fades, as the Potsdam joint dissolves and vanishes into the forgotten past. It coincides, in other words, with Tyrone's exit from the Allied housing compound. His departure marks the culmination of the totalitarian movement's accession to power. Having been partially stripped of his bourgeois identity and organized for the emerging order whilst entering the Zone, Tyrone next suffers the kaleidoscopic and chimerical images of the compound which constitute the movement's propaganda operation. His recovery of the hashish then marks the inception of totalitarian power, for having taken and tasted the bait he wanders carelessly into the Soviet sector of the compound and falls victim to agents of the new system: he is grabbed from behind by a snatch squad and injected with sodium amytal which leaves him 'clutching in terror to the dwindling white point of himself, in the first windrush of anaesthesia, hovering coyly over the pit of death' (*GR* 383). Tyrone is thus delivered into the new order of incipient totalitarianism, held at the edge of extinction in a state of permanent crisis, his tenuous identity suspended above those 'holes of oblivion' which Arendt calls 'the true central institutions of totalitarian organizational power.'[18]

The scene at Neubabelsburg dissolves as Slothrop loses consciousness. The compound joint, the complex boundary between imperialism and incipient totalitarianism, war and peace, history and myth, politics and entertainment, vanishes from sight and memory. A change in the structure of power entails a change in the structure of classification. Yet it is not the nationality of the snatch squad – a Russian team led by intelligence officer Vaslav Tchitcherine – nor the fact that Slothrop has

---

17. Cf. Arendt, p. 216.
18. Arendt, pp. 438, 459. Arendt is referring here to the concentration and extermination camps of the Nazi years; in many ways *Gravity's Rainbow* dramatizes the reabsorption of these camps into what will become the post-war world (see pp. 316-8, 488-9).

moved from the American to the Soviet sector of the compound that symbolizes the transformation. Rather it is his subsequent destination. Having entered Neubabelsburg through the rubble of Berlin's history in 1945, Slothrop leaves it by a very different route. For whilst he outmaneuvers the Soviet Army inside the compound Tyrone fails to spot Tchitcherine's police operation: when he regains consciousness he finds himself lying on an operating table inside a movie studio. Seduced by the Hollywood spectacle at Neubabelsburg, he now becomes part of its means of production. Politically, psychologically, socially, he absorbs and is absorbed by the new order (*GR* 383, 392).

Wandering through the studio Slothrop meets ex-film-actress Greta Erdmann who tells him that they are on the set of a movie called *Alpdrücken*, the first of 'dozens of vaguely pornographic horror movies' she starred in during the 1920s and 30s as 'the creature' of director Gerhardt von Göll. Her co-star was called Max Schlepzig. All of these films had the same plot: Greta was chased and chained by 'monsters, madmen, [and] criminals,' then tortured to exhaustion and death. Now she begs Tyrone to do the same and he obliges her. As he whips Greta the drives behind Slothrop's earlier Rocketman fantasy are realized in a penetrating dramatization of the pornography of power and a grotesque parody of the meaning of total individual license for any human community: both extract orgasmic pleasure from a ritual of extreme dominance and submission which edges towards mutual extinction.[19] The transition is thus complete. Having left the rubble of the imperial dissolution at Potsdam, Slothrop now enters the world of incipient totalitarianism through a partial re-enactment of a pre-Nazi German pornographic horror film. Having suffered personal terror at the hands of Tchitcherine's snatch squad, he now willingly participates in its routinization, enacting what appears to be a welcome program of thrills within a closed ideological environment, rehearsing a sequence that has run repeatedly – 'the same frame, over and over' – through the end of the Weimar Republic, interiorizing the commands of indoctrination from a mythic enclosure, and becoming a host of what Norman Mailer terms the totalitarian plague. Tyrone is coming over the rainbow, his re-entry caught in a frozen frame – 'our history is an aggregate of last moments' –

---

19. Cf. George Orwell, *Nineteen Eighty-four* (1949; Harmondsworth: Penguin, 1954), pp. 15-16.

into which Pynchon condenses the waves of the future: history dissolving into mythology, identity yielding to wilful obedience, the exercise of extreme power translated as endless entertainment, and the German surrender leading to a peace of compulsively mutual exploitation (*GR* 149, 392-7).

But this post-war environment is not a replica of pre-totalitarian Germany itself, nor has Slothrop simply been converted into a repressed, submissive 'little man,' the basic unit of Nazi power analyzed by Wilhelm Reich in *The Mass Psychology of Fascism* (1933) and other works.[20] Rather, his actions express Pynchon's sense of a more complex synthesis. When Greta tells him that her co-star was called Max Schlepzig, the name Tyrone carries on his identity card, he discovers that the disguises he adopted to enter Neubabelsburg have now adopted him and that he has been delivered into an unfamiliar world in which, under the director's hidden control, he is the central figure of attention. Slothrop therefore condenses the bourgeois dream of personal power into mass society. As long as he performs his role, his part in the script, and retains his alienated but politically reliable identity, he is as assured of fame, pleasure and sanctioned authority as Hollywood's most famous Tyrone. The world will fulfil his desires. In this sense Slothrop is reconstituted as the prototypical self of post-war advanced consumer capitalism: the descendant, as we have seen, of the isolated and lonely individuals who people the conclusion of Arendt's *Origins of Totalitarianism*, but at the same time the heir to the permissive codes of a society of *Führers*. As a result, his ability to act historically is under permanent and compulsive pressure, no longer from the totalitarian ideologies of race or class investigated by Arendt, but from the overtly tolerant and seductive opportunities of advanced capitalist society: the 'freedoms' which invite anyone to work, fight, consume, and spectate – in short, to behave – unfettered by social, religious, or traditional constraints.[21]

*Alpdrücken*, the film Greta and Slothrop rehearse, means nightmare, and its presence haunts the complex, interpenetrating

---

20. Wilhelm Reich, *The Mass Psychology of Fascism*, trans. Vincent R. Carfagno (1970; Harmondsworth: Penguin, 1975), pp. 17-18.

21. Arendt, pp. 473-9; Charles Olson, *The Special View of History*, ed. Ann Charters (Berkeley: Oyez, 1970), pp. 17-18; Hannah Arendt, *The Human Condition* (Chicago: Chicago University Press, 1958), pp. 38-49; Herbert Marcuse, 'Repressive Tolerance,' in Paul Connerton (ed.), *Critical Sociology* (Harmondsworth: Penguin, 1976), pp. 301-29.

spaces of *Gravity's Rainbow*, extending directly into the historical field to influence both sexual fertility and Nazi politics. Pynchon uses it as an integration sign to conflate moments of power – the erotic, political, and mythological – conventionally defined as distinct. As such it constitutes a crucial intersection point in the production of the incipient totalitarian structure which matures and expands beyond the Allied housing compound, and an expression of the new environment Geli Tripping earlier announces at Nordhausen, where the German V–1 and V–2 rockets had been assembled. There she tells Slothrop: 'Forget frontiers now. Forget subdivisions. There aren't any ... it's all been suspended ... there are no zones. No zones but the Zone.'[22] To this lesson Pynchon adds a direct comment which restates Geli's advice in terms of twentieth-century physics and stresses the mutability of classification: 'here in the Zone categories have been blurred badly ... here are only wrappings left in the light, in the dark: images of the Uncertainty.' (*GR* 294, 303, 333, 394-7, 429)[23]

The word 'zone,' with its erogenous connotation, suggests an exploitable bank, and Pynchon investigates definitions of unity and category as tools of power. The *Alpdrücken* film set therefore constitutes a second, ascendant element in Pynchon's drama of the restructuring of power beyond the imperial order. It was built in the first years of the transformation, an early manifestation of Neubabelsburg's influence; it survived the armed conflict which marked incipient totalitarianism's accelerated growth; and in 1945 it stands as Pynchon's image of what is to come post-war. The film's title is hardly coincidental (Greta tells Tyrone that von Göll shot it 'at the height of his symbolist period'): in this commentary on Stephen Dedalus's line from James Joyce's *Ulysses* (1922) – 'history is a nightmare from which I am trying to awake' – Pynchon portrays Slothrop waking from the history of World War II concluded at Potsdam, not into the freedom that conference delegates insist will accompany the defeat of Nazism, but into a nightmare of terrifying thrills which Pynchon sees encircling the post-war world (*GR* 394).[24]

---

22. Cf. Herman Melville, *Moby Dick; or The Whale*, ed. Harold Beaver (Harmondsworth: Penguin, 1972), pp. 227-8.

23. Robert Jay Lifton, *Boundaries: Psychological Man in Revolution* (New York: Vintage, 1970), gives the context.

24. James Joyce, *Ulysses* (1922; Harmondsworth: Penguin, 1969), p. 40; Clerc in Clerc (ed.), p. 118.

## Concord in a State of Arms

Well, that changes warfare!

— Douglas MacArthur (August 1945)

From within the boundaries of this analysis of the restructuring of power Pynchon decodes the history of World War II and its conclusion: the destruction of Nazi Germany at the hands of the Grand Alliance. Consistent with his sense of the war as the central phase of a larger transformation within an emerging global system of power, Pynchon analyzes the victory of the Allies in 1945 not simply as a conclusion but also as a revelation. Historically, the collapse of the Third Reich brought with it the end of large-scale national military conflicts in the west; but in the process it revealed the emergence, accelerated by the war, of new forms of conflict which incorporated characteristics of and expressed the survival of totalitarianism beyond the Nazi capitulation. These struggles displaced the now unnecessary and counterproductive wars of the imperial era to the margins of their previous domain and consequently intensified inside the remaining shells of military conflict.[25] Their natures reflected the legacy of the war between the two sides, moving from open armored confrontations toward covert mass propaganda operations, from war between nation-states toward the non-military 'pacification' of a multinational population by authorities who, beyond the officially maintained conflicts, shared common methods and goals.

These new struggles are anticipated at Neubabelsburg. Slothrop, an American, wanders into the Soviet sector of the Allied housing compound and is drugged by a Russian snatch squad who subsequently process him and abandon him on a movie set. He appears to be an early victim of the Cold War. But nationality is of little significance here because the sodium amytal used by Tchitcherine's team to knock out Slothrop has also been used by the Anglo-American Abreaction Research Facility team to drug him earlier in London. There he had officially been a volunteer; here he is taken by surprise. The results are the same. In both

---

25. Military conflict and the use of armed force do not, of course, disappear but are moved away from the metropolitan centers of the industrialized west to their peripheral regions. See Fred Halliday, *The Making of the Second Cold War* (London: Verso, 1983), p. 81.

cases he is manipulated by state authorities using identical methods for their own reasons, and on neither occasion is he told why. Like a latter-day Lemuel Pitkin, the victimized hero of Nathaniel West's *A Cool Million* (1934), Tyrone is not killed in battle but rendered senseless with narcotics administered by experts. Having been drugged, experimented on, and redeployed by the authorities, he then finds himself the ostensibly innocent agent of power. On the *Alpdrücken* film set the scene dissolves as Tyrone, an American put there by the Russians, takes on the role of a German Jew, Max Schlepzig, to punish Greta, a German actress who had entertained the troops during the war. Even as they justify their presence by reference to 'the other side,' the victorious powers (disguised as local victims) are coming together literally to whip the losers into line. But in keeping with the changing nature of conflict, there is no overt force involved. As Dr. Benway remarks in William Burroughs' *The Naked Lunch*, 'It's not efficient.'[26] More important, since Greta begs Slothrop to whip her, it is no longer necessary (*GR* 61, 114, 383, 390).

Pynchon's analysis of this dual action reflects his reading of the nature of the Cold War. Just as in the novelist's history of *Gravity's Rainbow* the defeat of Nazi Germany spawns a struggle between the victors for the spoils which is in turn dramatized as a synthesis within the larger development of incipient totalitarianism, so in the history of the years after 1945 the emergence of the Cold War may be interpreted not simply as a tragic conflict between powers but also as a form of international organization required inside the post-imperial structure. In this sense the Cold War appears as a unified system characterized by the presence within it of controlled divisions: in Pynchon's words, 'concord in a state of arms' (*GR* 748).

With the German collapse the reorganization became visible. Soviet and American troops first linked up at the River Elbe on 25 April; five days later Hitler committed suicide, and on 7 May Jodl and von Friedeburg signed the unconditional surrender. But the Allies' celebrations were shortlived, for if the frontiers and subdivisions were suspended, as Geli Tripping told Slothrop in *Gravity's Rainbow*, new ones were soon to be imposed. Three months later at Potsdam, Truman, Churchill, and Stalin bargained closely over the spoils of victory and the new order in Germany and eastern Europe. Disagreements there, even when

---

26. William Burroughs, *The Naked Lunch* (1959; London: Corgi, 1974), p. 39.

bypassed or resolved in the short term, presaged crises in Iran and Turkey; Potsdam anticipated the restoration of conflict. As Daniel Yergin notes, 'because they could not agree on how to govern Europe, they would begin to divide it.'[27] The Cold War beckoned. And yet when Geli Tripping told Slothrop to 'forget subdivisions' she hinted at an implicit, though vital, agreement between the victorious powers. The emerging partition of Europe also expressed the continuing necessities of state power in east and west which united the former Allies after the collapse of their common enemy: 'on each of the outstanding issues the Great Powers found that the best way to cooperate was to give each a freer hand in its own sphere.' They had a common interest in milking the loser. By July 1945, that 'freer hand' was already stripping material from all zones inside the Zone.[28]

Here, as elsewhere, Pynchon's fiction articulates one of the definitive characteristics of the occupation. Reparations may not have been able to replace the twenty million Russians who perished in the conflict, but between 1945 and 1946 Soviet forces did systematically strip the areas they had occupied of everything from heavy industrial equipment and railroad rolling stock to agricultural machinery and livestock, from house furnishings and plumbing fixtures to the filing cabinets in Hitler's bunker, in order to help them repair the damage done to the nation's fabric as a result of the fighting. In the western sectors, former US Ambassador to the Soviet Union Joseph E. Davies reported that the French too were 'carrying everything, including the kitchen stove, out of their territory,' and added that both American soldiers and members of the US reparations delegation led by Edwin Pauley were 'liberating' items from within their zone. In rocket equipment alone, American officials removed some 400 tons of hardware and 14 tons of printed matter (including thousands of technical reports and over half-a-million engineering drawings) to the Aberdeen Proving Ground, Maryland, before the end of 1945. When questions concerning the nature, scale, and primacy of reparations payments led to disagreements between the former Allies and to a suspension of shipments from the western to the

---

27. Yergin, p. 118.
28. Yergin, pp. 7, 118. See also Charles L. Mee, *Meeting at Potsdam* (New York: Evans, 1975), pp. xiii-xiv; John Lewis Gaddis, *The United States and the Origins of the Cold War, 1941-1947* (New York: Columbia University Press, 1972), pp. 242-3; Walter Lippmann, *U.S. Foreign Policy: Shield of the Republic* (London: Hamish Hamilton, 1943), pp. 72, 103-5.

eastern zones in May 1946, the *de facto* military division of Germany started to take on economic and political dimensions that would lead through the creation of a unified Anglo-American area of occupation ('Bizonia') and proposed currency reforms in the west to the Berlin blockade and the eventual creation of two separate German states.[29]

What survived across these burgeoning subdivisions is one of the central concerns of *Gravity's Rainbow*: 'the rationalized power ritual that will be the coming peace' (*GR* 177). That power ritual, prefigured at Potsdam, closed World War II, but it did not terminate warfare so much as inherit and restructure it. As Lewis Mumford argues, the collapse of Nazi Germany at the hands of the Grand Alliance in 1945 constituted a transformation inside the larger development of what he terms 'the modern megamachine,' not the extermination of some irrational isolated evil: 'by the curious dialectic of history, Hitler's enlargement and refurbishment of the Nazi megamachine gave rise to the conditions for creating those counter-instruments that would conquer it.'[30] Even in defeat, the influence of the Axis persisted through the new synthesis: Nazism prepared the ground for the emergence of a system which no longer needed it. What surfaced in 1945 was an incipient totalitarianism, a structure of power which inherited characteristics from its immediate past and from all sides in the conflict, but which restructured them in the light of its new potentials, limits, and necessities. And since war remained, in Mumford's words, 'body and soul of the megamachine,' what survived in victory and defeat was the necessity of conflict: 'in order to keep [the] megamachine in effective operation once the immediate military emergency was over, a permanent state of war became the condition for its survival and further expansion.'[31]

That state of war took its initial outward character from the cease-fire lines of World War II with the division of Germany into

29. Yergin, p. 64; Gaddis, pp. 239, 242-3, 325-32. See also John Gimbel, *The American Occupation of Germany: Politics and the Military, 1945-1949* (Stanford: Stanford University Press, 1968); J.P. Nettl, *The Eastern Zone and Soviet Policy in Germany 1945-1950* (London: Oxford University Press, 1951); Robert M. Slusser (ed.), *Soviet Economic Policy in Postwar Germany* (New York: Research Program on the USSR, 1973).

30. Lewis Mumford, *The Myth of the Machine, Vol. II. The Pentagon of Power* (London: Secker & Warburg, 1971), pp. 233, 250-52. Cf. Norman Mailer, *The Naked and the Dead* (1948; St. Albans: Panther, 1964), p. 274.

31. Mumford, p. 256.

four administrative units. But following the cessation of hostilities, American political, military, and economic stakes in the division of Europe mounted. In order to re-establish the open world economy that policy makers believed essential to US prosperity and global stability, to finance the post-war US export surplus, to restore western European production and competitiveness and stabilize their currencies, to undercut the left's appeal and the extension of the closed economies of the nascent Soviet bloc, and to help finance renewed imperial wars on the periphery, the British loan (1946) and the larger Marshall Plan (1947-51) poured billions of dollars into western Europe. The existence of a Soviet 'threat' to the region played a vital role in the passage of both programs, with the British withdrawal from Greece (announced in February 1947), the Czech coup (February 1948), and the blockade of Berlin (June 1948 to May 1949) providing the Truman administration with evidence to justify these and other initiatives such as the Truman Doctrine (March 1947), the creation of NATO (April 1949), and a major expansion of the Strategic Air Command. The outbreak of the Korean war in June 1950, following the communist victory in China and the explosion of the first Soviet atomic bomb, provided an equally opportune justification for the introduction of the international rearmament program (proposed by NSC–68 in March 1950), which by increasing US 'defense' spending from $13.5 billion to $40 billion per annum pumped more dollars into Europe to complete the unfinished work of the near-terminated Marshall Plan.[32]

Hence the economic, political, and social requirements of an increasingly international system came to depend on the maintenance of a plausible military threat, initially in Europe and eventually across the globe. And the existence of that threat, real or manufactured, in turn bolstered the growth of the central institutions of the system, the nascent world government of multi-national corporations and banks.[33] In the wake of the Marshall

---

32. Fred L. Block, *The Origins of International Economic Disorder* (Berkeley: University of California Press, 1977), pp. 32-108; Fred L. Block, 'Economic Instability and Military Strength: The Paradoxes of the 1950 Rearmament Decision,' *Politics and Society* 10 (1980), pp. 35-58; Richard M. Freeland, *The Truman Doctrine and the Origins of McCarthyism* (New York: Alfred A. Knopf, 1972), pp. 66-8, 82-101, 151-201, 234-87; Yergin, pp. 178, 279-84, 303-5; Gaddis, pp. 342-3, 348-52.

33. Alan Wolfe, *The Rise and Fall of the Soviet Threat* (Washington: Institute for Policy Studies, 1979); Harry Magdoff, *Imperialism* (New York: Monthly Review Press, 1978), p. 173.

Plan and other American initiatives – the Bretton Woods banking system, International Monetary Fund, World Bank – the internationalization of the US-dominated business system gathered pace. The total value of American manufactured exports and sales by overseas-based American firms increased from $15.8 billion in 1950 to $39.7 billion in 1960. Of greater significance, the book value of direct US foreign investments abroad also grew: in Europe from $1.73 billion in 1950 to $6.69 billion in 1960 and $24.52 billion in 1970; globally, over the same period, from $11.79 billion to $31.82 billion and $78.18 billion. By the early 1970s 52 per cent of the world's foreign investments were in the hands of American-based multinational corporations and banks, the result of an eight-fold increase in overseas capital accumulation by US firms since 1945. Such developments depended on the creation of an open world economy and the pursuit of the Cold War.[34]

The symbiotic tensions in this relationship are active in Pynchon's novel. At Nordhausen, for example, the discussion between Tyrone Slothrop and Geli Tripping gives access to both the continuities and transformations inside the modern power structure which are evident as World War II is 'adjourned and reconstituted as a peace.' When Geli tells Slothrop to ·forget frontiers and subdivisions because 'there aren't any,' Tyrone replies 'there are soldiers.' 'That's right,' says Geli, 'but that's different.' Here each connotes dialectically related moments within the processes of advanced capitalism whose dream of global power requires conflict as its method. The frontierless Zone which Geli proposes anticipates the growing necessity of multinational banks, corporations, and communications systems: the destruction of the British Imperial Preference system and of other barter and cartel arrangements, the reduction of tariff barriers, and the elimination of exchange controls and currency inconvertibility, all of which restricted the access of American firms to overseas trade and investment opportunities by preserving the closed economic blocs which so damaged US prosperity in the 1930s.[35] The soldiers that Slothrop has to negotiate en route to Neubabelsburg embody one major inheritance from the struggle

34. Harry Magdoff, *The Age of Imperialism* (New York: Monthly Review Press, 1969), p. 180; Mira Wilkins, *The Maturing of Multinational Enterprise: American Business Abroad from 1914 to 1970* (Cambridge, Mass.: Harvard University Press, 1974), p. 330; Magdoff, *Imperialism*, p. 170.

just resolved. But they are an inheritance which was soon to be displaced, with the successful testing of the world's first atomic device by scientists at Alamogordo, New Mexico, during the Potsdam conference on 16 July. The crucial question raised in *Gravity's Rainbow* – 'after the act of wounding, breaking, what's to become of the little Oven State?' – is already being answered: the Oven State, 'salvaged ... patched up, roles reassigned,' relieved of the paradoxes and inefficiencies that undermined its previous incarnation, will take on a new, more appropriate form: 'the War itself as tyrant king' (*GR*, 75, 102, 294).

In July 1945 the interrelations between armed conflict and the new demilitarized Zone are dramatized in a scene which Pynchon composes as a preview of what is to come 'post-war.' Stumbling through the ruins of Berlin, Slothrop wonders what has happened to the city of military order he knew from newsreels and magazines. Pynchon's image of early post-war Berlin suggests that the priorities which shaped Hitler's capital persist inside the ruins of defeat:

> If there is such a thing as the City sacramental, the city as outward and visible sign of inward and spiritual illness or health, then there may have been, even here, some continuity of sacrament, through the terrible surface of May. The emptiness of Berlin this morning is an inverse mapping of the white and geometric capital before the destruction ... everything's been turned inside out. The straight-ruled boulevards built to be marched along are now winding pathways through the waste-piles, their shapes organic now, responding, like goat trails, to laws of least discomfort. The civilians are outside now, the uniforms inside. (*GR* 372-3)

Behind 'the terrible surface of May' and Germany's military surrender the totalitarian dream remains. Having discovered what Arendt defines as 'a means of dominating and terrorizing human beings from within' (one whose consequences were already being analyzed through Wilhelm Reich's concept of 'character armor'), the post-war system inherits and relocates overt militarism inside the new folds of civilian adjustment; an exhausted population surfaces to accept a deepening occupation:

---

35. Gabriel Kolko, *The Politics of War* (1968; New York: Vintage, 1970), pp. 242-340; Block, *Origins*, pp. 32-137; Laurence H. Shoup and William Minter, *Imperial Brain Trust* (New York: Monthly Review Press, 1977), pp. 117-87; Block, 'Economic Instability,' pp. 37-8.

'Inside is outside ... outside has been brought inside.' So that whilst Slothrop has to evade soldiers to enter the Allied housing compound, they are present as detachments from World War II, figures of military inheritance guarding a temporary conference. Geli Tripping's frontierless civilian Zone more accurately figures the grounds of future conflict and the shape of war to come.[36]

## Tyrone Slothrop: The Wave of the Future

'To the babies.' Grinning, completely mad.
'Babies, Gwenhidwy?'
'Ah. I've been keep-ing my *own* map? Plot-ting da-ta from the maternity wards. The babies born during this blitz are al-so fol-lowing a Poisson distribution, you see.'
'Well – to the oddness of it, then. Poor little bastards.'

— Thomas Pynchon, *Gravity's Rainbow* (1973)

Tyrone Slothrop is, as we have seen, en route through that Zone, and it is mainly by dislodging Slothrop from his established imperial role inside the established imperial order that Pynchon accumulates and displays a field of events, relationships, and data which elaborate the changing nature of power and the intricate structure of conflict the occupation expresses. Tyrone is on the run, having slipped World War II and its imperial network by going absent without leave from his job with the Political Warfare Executive (PWE), which he had been assigned to from his desk in London at ACHTUNG (Allied Clearing House, Technical Units, Northern Germany) and which had taken him from the Abreaction Research Facility (ARF) at Saint Veronica's Hospital on the other side of the city to the Casino Hermann Goering on the French Riviera. The ARF, an appropriately named center for darkly comical Pavlovian animal trainers, gives him his first taste of the sodium amytal which will later ease him away from the Allied housing compound at the Potsdam conference. Tyrone quits the casino due to a mixture of unfocussed but growing

---

36. Cf. Arendt, *Origins*, p. 325; Wilhelm Reich, *The Discovery of the Orgone, Vol. 1. The Function of the Orgasm*, 2nd edn, trans. Vincent R. Carfagno (New York: Simon & Schuster, 1973), pp. 7-8, 138-47; Wilhelm Reich, *Character Analysis*, 3rd, enlarged edn, trans. Theodore P. Wolfe (1949; New York: Farrar, Straus, & Giroux, 1970), pp. xxiv-xxv, 39-44.

paranoia born of strange coincidences and the disturbing behavior of some of his colleagues, a desire for 'freedom', and a quest for information. His discovery of the mysterious 'Imipolex-G' while still at the casino leads to his meandering journey into the Zone in search of further information to elucidate his vague sense of a concealed structure of malevolent power focussed on him. The structure is in some way connected with the production of German V–2 rockets, one of which incorporates an insulation device made of Imipolex-G. The search for this substance and the device itself, the 'S-Gerät,' leads Slothrop through a series of contacts — targets Laszlo Jamf and Franz Pökler, informants Blodgett Waxwing and Mario Schweitar – via Lyle Bland, then back to his own father Broderick and his childhood. Other trails cut across nations and back through time, linking Slothrop's upbringing into a vast network of obscurely related plots. As he accumulates information Tyrone decodes the limits of his freedom, which he has increasingly suspected as the map of his assemblage has unfolded (*GR* 185 ff.).

But Slothrop is not simply a neutral data gatherer beyond any historical or social context, nor is Pynchon's handling of the changing nature of warfare restricted to a McLuhanesque reading of conflict as 'an involuntary quest for identity.'[37] In his journey into the Zone Slothrop illuminates the whole field of forces his requirements activate. The search for information intersects with his fear of disclosure and need of food, sex, and shelter inside a mobile network of codes and decodings, contacts known and unknown, chance meetings, warnings, and advice. Having fled his PWE work on the French Riviera, Tyrone needs alternative sources of money to help feed himself, finance his travels, and purchase data. A grant from Blodgett Waxwing's associate in Nice gets him to Zurich, where he pays for the Sandoz Drug Company's confidential files on Laszlo Jamf and the Imipolex-G he synthesized ('bootlegged by Mario Schweitar') by working as a courier for Francisco Squalidozzi. Information transferred and received is therefore related to employment: Tyrone works in order to consume and view. But his needs draw him into hypnotic diversions. In Berlin, 'beguiled' by the prospect of a million German marks and a young lady called Magda, Slothrop agrees to act as a recovery man for Seaman Bodine. This is how

---

37. Marshall McLuhan, Quentin Fiore, and Jerome Agel, *War and Peace in the Global Village* (New York: Bantam, 1968), p. 123.

he comes to be at the Allied housing compound in Neubabels-burg. In taking the job, which rescues him from having to steal food, Slothrop's aims are necessarily mediated: the search for data requires recovery of the hashish. The payment is itself narcotic rather than directly informative: a kilo of dope and a million counterfeit marks. The whole deal is arranged under the influence of Emil Bummer's reefers. In the midst of the Zone the quest for information becomes chimerical, and Slothrop discovers that he is in a position similar to that of Yossarian in Joseph Heller's *Catch-22* (1961): in order to decode the accumulating clues and hence to reveal the hidden order he is obliged to perform tasks which at the same time serve to conceal it. The very act of clarifying the concealed territory requires a blurring of his own map so that as he approaches his target he also avoids it. Tyrone finds himself part of Heisenberg's uncertain world: he is involved (*GR* 256-371).

This episode from the middle of *Gravity's Rainbow* indicates one of its prime concerns: an analysis of the hallucinatory quality of information inside complex systems. As World War II reaches its conclusion, conflict is reconstituted towards a civilian struggle over the 'universe of images' which govern behavior through education, employment, the media, and inherited myths of social process. Slothrop's ability to outwit the Soviet security guards instances the succession: conflict is transformed from the inefficient military operations of the battlefield to the profitable manipulation of identity and image on any territory; the illusionist replaces the soldier. In Berlin and Neubabelsburg Pynchon dramatizes the expanding field of warfare within an increasingly propagandized post-war culture: the confused definitions of life and fantasy which result from the perpetual avalanche of purported fact on any society untrained for an analysis of the nature of power; the points at which understanding yields to hallucination inside the compulsive governing mythologies of advanced capitalist societies. Pynchon's Zone thus anticipates the conditions of post-war American life in yet another way: it figures the structures of power as increasingly predicated on the control of communication and the manipulation of images.[38] At the same time Tyrone Slothrop embodies the emerging unit of that coercive society of both Left and Right which spreads across the

---

38. Cf. Jacques Ellul, *Propaganda: the Formation of Men's Attitudes*, trans. Konrad Keller and Jean Lerner (1968; New York: Vintage, 1973), pp. 84-7, 112-16.

post-Potsdam world: an engineered and engineering figure on perpetual patrol, acquiring and discarding identities as he needs them, constituting and reconstituting himself in relation to the shifting and uncertain environment of the Zone, acting and reacting ambivalently inside obscure limits. He points towards a state of pervasive struggle in which the clash of armies is supplanted by the anxious adjustment of the enlarged forces, the soldier–worker–consumers of advanced capitalism.[39]

But if Slothrop is, as one of his many contacts puts it, 'the wave of the future' his existence is a function of his motion under external forces and his prospect is limited: he will break up on the shores of the post-war world. In the second half of *Gravity's Rainbow* Slothrop's laminated character is gradually eroded under the changing conditions and demands of 'peacetime.' Near the end of the novel Pynchon recalls a story about how Tyrone 'was sent into the Zone to be present at his own assembly' but how now he 'is being broken down instead, and scattered' (*GR* 258, 738). The dismantling expresses the requirements of the 'Operation'. Slothrop's autonomy is restricted by the needs of the restructured war and its agents, and the basis of his identity is emasculated in the interests of the power structure:

> The man has a branch office in each of our brains, his corporate emblem is a white albatross, each local rep has a cover known as the Ego, and their mission in this world is Bad Shit ... Tyrone ... has become one plucked albatross. Plucked, hell – *stripped*. Scattered all over the Zone. It's doubtful if he can ever be found again, in the conventional sense of 'positively identified and detained.' (*GR* 712-13)

Conflating materials from Arendt and Reich, from Melville's *Moby Dick* and Coleridge's *Rime of the Ancient Mariner*, Pynchon dramatizes the manipulation of identity through a series of interlocking metaphors. The albatross is presented as an emblem of benevolent power – 'hailed ... in God's name' as 'a Christian soul' in *The Ancient Mariner* (I. 65-6) – and thus as the agent of a necessary salvation from the feared world of nature and natural man. But once accepted as such through an interiorization of command given in appropriate commercial and bureaucratic terms, its motives are revealed. Tyrone, himself 'a Christian soul'

---

39. See Lifton, pp. 37-63.

of Puritan descent, suffers the fate of the albatross and becomes
the sacrificial victim of power inside a ritual drama of resurrection
and hence its redeemer: a Hanged Man whose baptism involves
his own death; an impersonal container and projector of The
Man's energy, caught and scattered in the wind. Rendered
impotent and isolated, Tyrone's directed power on the *Alpdrücken*
film set now appears as the alienated labor of any employee
within a hierarchy of control. Like the carpenter in *Moby Dick* – 'a
stripped abstract ... unreasoning but still highly useful' – Slothrop becomes nothing more than a technical implement, a 'tool by
which the ends of power are achieved.' Reduced to a 'transmitting mechanism in the machinery of state,' he is left, like the
carpenter, as 'pure manipulator' bereft of political identity: the
ideal unit for a world experiencing the dissolution of the public or
political realm.[40]

Yet Slothrop does not simply vanish into the sands of time. In
his character Pynchon condenses what he sees as the various
forms of social life sanctioned by the emerging division of behavior, so that Slothrop is presented not only as a wave but also as a
particle whose decay releases new elements. Just as Nazism
prepared the conditions for the development of a system which
no longer needed it, so Tyrone carries the seeds of identities

---

40. Cf. Clive Bush, *The Dream of Reason* (London: Edward Arnold, 1977), p.
55; Eric Mottram, 'Performance: Charles Olson's Rebirth between Power and
Love,' *Sixpack* 6 (1973/74), pp. 96-7; Melville, *Moby Dick*, p. 579. The circumstances of Slothrop's apparent isolation and entrapment as such in the Zone are
indicative of his transitional historical status. Seeking a route out of the Zone and
back to the United States, he approaches Gerhardt von Göll for those credentials
he believes necessary to secure his evacuation alongside the V–2 rockets being
tested at Cuxhaven or prepared for transfer to New Mexico. Von Göll in turn
proposes furnishing the documents at Putzi's, a popular entertainment center in
Cuxhaven which itself appears to Slothrop on his arrival to be an escape valve from
warfare and the Zone: a border post 'seal[ed] off completely from the outside' in
which an absorbing and interminable 'victory party' accommodates all-comers; a
switching point giving access to the Rocket State's own 'enormous transit system'
beyond. However, what Slothrop discovers en route to Putzi's is that accession to
post-war involves more than the mere doffing of khaki. Directed along his own
'yellow brick road' by one self-promoting narcotics dealer, Albert Krypton, and
then taken in at Putzi's only to be strung out between others, like Seaman Bodine,
Slothrop finds that safe passage beyond the imperial order requires his own
thorough post-imperial reassembling. As Pynchon notes, and as von Göll's initial
mockery and subsequent failure to show up confirm, this sort of forging 'take[s]
time.' Slothrop's subsequent dissolution is therefore only the beginning of a
process – which Tyrone describes as 'some kind of a plot' – being carried out in
various guises right across the Zone. See *Gravity's Rainbow*, pp. 526-7, 594-603.

which will sprout and flourish in the advanced capitalist environment beyond Potsdam. In almost the final reference to him,
Pynchon notes how the shell-like layers of Tyrone's identity
survive down the years: 'Some believe that fragments of Slothrop
have grown into consistent personae of their own. If so, there's no
telling which of the Zone's present-day population are offshoots
of his original scattering' (*GR* 742). To this extent Tyrone is a
mosaic preview of the future and an agent of historical development. The remolding and processing he undergoes as he enters
the Zone constitute a capital investment program designed to
ensure the system's survival beyond its imperial phase. He
becomes, in effect, part of its means of reproduction and adaptation, an experimental identity unit on whom man's future
selves are tested and proven. The return on that investment
comes with Tyrone's programmed decay – his withdrawal from
circulation – and the launch of the new products onto the
markets of the post-war world: the offspring who have no
recorded father beyond 'The Man' inside their brains.[41]

Even in the absence of certainty, it remains statistically
probable that such decay favors stability – or, more precisely, the
desire for security – and increases entropy. Tyrone's disarming
and fragmentation yields increasingly anonymous personnel
units adapted to the needs of the Operation. His offshoots, in Eric
Bentley's terms, carry post-war society towards morphological as
well as functional differentiation. Slothrop himself comes to rest
as an ahistorical nucleus similarly employed: he is last seen as a
photograph on the cover of a record album by an English rock
group called The Fool (*GR* 742). He therefore becomes a tarot
figure of risk, potential, and ecstatic innocence used in the world
of mass production marketing; a fertile anarchist at the service of
a music industry whose denial of social limits serves only to
extend them. Slothrop escapes location and structure at the cost
of identity and autonomy. In each case his division increases the
entropic chaos of oversystemized energies as, in Hannah Arendt's
sense, public *praxis* is displaced by social behavior.[42]

---

41. See Herbert Marcuse, *Eros and Civilization* (1955; Boston: Beacon Press,
1974), pp. 96-101.
42. Marshall McLuhan, *Understanding Media* (1964; London: Abacus, 1973), p.
80; Eric Bentley, *The Cult of the Superman* (London: Robert Hale, 1947), p. 236;
Arendt, *Human Condition*, Part II, pp. 38-49. My use of Arendt is in the context
provided by H.T. Wilson, *The American Ideology: Science, Technology and Organization
as Modes of Rationality in Advanced Industrial Societies* (London: Routledge & Kegan
Paul, 1977), pp. 63-5, 171-84, 197.

The products of this process are the victims of the war beyond Potsdam, the pervasive conflict at the core of advanced capitalism. Whereas the military campaigns of World War II entailed physical death through bombs and bullets, peacetime brings its own casualties and fatalities, however camouflaged. As 'Mister Information' recalls in Autumn 1945:

> the Germans-and-Japs story was only one, rather surrealistic version of the real war. The real war is always there. The dying tapers off now and then, but the War is still killing lots and lots of people. Only right now it is killing them in more subtle ways. Often in ways that are too complicated, even for us, at this level, to trace. (*GR* 645)[43]

Tyrone's fate anticipates that of his descendants: he will be neither clear winner nor obvious victim. Just as the victory or death of the battlefields of World War II are replaced by the unresolved tension between anxiety and security characteristic of the Cold War, so Slothrop looks forward to a future of tense and inconclusive endurance. His tarot cards 'point only to a long and scuffling future, to mediocrity ... to no clear happiness or redeeming cataclysm.' All signs of hope are reversed: the future guarantees no deliverance, only polymerization. The comments of the late German Foreign Minister Walter Rathenau on the meaning of chemical synthesis apply equally to Tyrone and the continuing war: 'the real movement is not from death to any rebirth. It is from death to death transfigured.' And if Slothrop's former shells lie scattered across the Zone, waiting to proliferate into the officers and men of the reconstituted war, his emasculated core ends up with its casualties: framed and unidentified amidst the rubble of war, 'among the humility, among the gray and preterite souls,' the crippled victims of a power structure which continues to demand sacrifice through war and peace (*GR* 166, 738, 742).

## Edward Pointsman: Towards the Sober Shore

'You see, you see,' the voice starts up again ... 'How many chances does one get to *be* a synthesis, Pointsman? East and West, together

---

43. Cf. Norman Mailer, *The Presidential Papers* (1963; St. Albans: Panther, 1976), p. 142.

in the same bloke? You can not only be Nayland Smith, giving a young lad in a funk wholesome advice about the virtues of work, but you also, at the same time, get to be *Fu Manchu!* eh? the one who has the young lady in his power! How's *that?* Protagonist, and antagonist in one. I'd jump at it, if I were you.'

— Thomas Pynchon, *Gravity's Rainbow* (1973)

Through Slothrop, Pynchon presents the restructuring of identity under the pressure of capitalism as it absorbs World War II and settles into peacetime. His fragmentation dramatizes the erosion of personal power inside a structure of pervasive conflict where politics, economics, and the manipulation of social psychology and information interact in the interests of elite rule and its extension. Slothrop is broken down and removed from history: his engineered shells fill the naturalized environment of post-war morphological differentiation whilst his anarchic core is displaced to an ahistorical world of fantasy beyond; his political identity dissolves into the pre-political worlds of familial hierarchy and of nature itself. But Slothrop does not decay in a vacuum, and Pynchon's analysis necessarily includes the agents of his scattering whose operations provide a history of Tyrone's dehistoricization. Their presence emerges as Slothrop quits the imperial order and is drawn into the world of incipient totalitarianism. For although Slothrop does not know it, his decision to escape from the Casino Hermann Goering is not solely a function of personal will following the reported death of his friend Tantivy Mucker–Maffick, but depends on the decisions of others whose changing financial, technical, and informational capabilities illuminate the processes of power evolving through and beyond V–E Day. Initially, Tyrone is offered assistance by Blodgett Waxwing, but Slothrop knows nothing of his involvement in the conspiratorial system he senses closing in on him. Waxwing is, as his mythical surname suggests, a forger and contrabandist: he is on the run from a military stockade. Like Daedalus, he furnishes Tyrone with a means of liberation, the wings in this case consisting of money and contacts. But, as in the myth, those wings prove unreliable when tested near the center of energy. Where Icarus approaches the sun, loses his wings, and crashes into the Aegean Sea, Slothrop enters the Allied housing compound and approaches the White House, loses his cash crop (the hashish) and contact with the outside world when drugged, and falls into the depths of totalitarianism. More significantly, Tyrone is *allowed* to escape from the casino by Dr. Edward Pointsman, the effective controller of 'The White Visitation' (which had transferred Slothrop from ACHTUNG to the Abreaction Research Facility and the Riviera) and coordinator of his surveillance (*GR* 248, 256-7, 270).

Pointsman acts as the essential vehicle of the transformations in the nature of control across the boundary of World War II; his movements in the world of power parallel those of Slothrop in the Zone's laboratory of victims. He takes up the Operation's designs at the point where the concentration camp system of the Third Reich, having produced what Arendt describes as its legions of 'ghostly marionettes with human faces,' can go no further. Pointsman's dream of completing Pavlov's unfinished research on the psychological bases of behavior, which is focussed through his experiments on Slothrop, entails an extension of control towards the behavioral programming of humans inside a predictable causal environment; an absorption of Pynchon's Zone within some as yet undefined steady state. He anticipates the engineered world proposed in B.F. Skinner's contemporaneous *Walden Two*.[44] By manipulating into dependence the official director of 'The White Visitation,' an aging, World War I vintage, traditional British military figure named (appropriately) Brigadier Pudding, Pointsman advances his own post-military program of control. But he knows that his work depends on continued funding from the PWE and that, with the cessation of hostilities in sight, his 'moment's stewardship' at the organization is in danger. Determined to survive and to carry his plans into the post-war world, 'facing squarely the extinction of his program,' Pointsman casts around for more long-term sources of finance and enlists the support of Imperial Chemicals through one of their plastics experts, Clive Mossmoon, and of Shell–Mex via Dennis Joint. He therefore shifts financial control of his social engineering program from the military to the expanding international business system in the interests of its long-term success. As 'Mister Information' says, he is quite literally 'the pointsman ... because he throws the lever that changes the points' which are located in the war. He directs the train of operations away from the 'Pain City' of outmoded militarism, which it had been fleeing

---

44. B.F. Skinner, *Walden Two* (1948; New York: Macmillan, 1976). As Skinner notes in his 1976 introduction, the book was actually written during the early summer of 1945. See also Arendt, *Origins*, p. 455.

at the start of *Gravity's Rainbow*, and towards the 'Happyville' of an engineered utopia (*GR* 75-6, 89-90, 227-8, 644-5).

But this survival plan rebounds on Pointsman. Just as 'the wave of the future' Tyrone Slothrop breaks up on the shores of post-war so Pointsman throws the switch only to be left at the junction. After discussions with his new financiers Pointsman decides to let Slothrop escape from the Casino Hermann Goering, in one move granting Tyrone the illusion of freedom while at the same time cutting the costs of continued surveillance by reassigning the task from PISCES to the Secret Service.[45] But the latter lose track of Slothrop, thus precipitating a crisis for Pointsman, who has long dreaded the loss of his key subject, as well as for Shell–Mex, who fear that the Russians might capture him and extract details of British V–2 rocket intelligence. Exacerbated as it is by growing internal and external pressure on PISCES due to the ailing Brigadier Pudding's shortcomings and the end of the war, the shock of Slothrop's escape effectively reveals the drive behind Pointsman's strategy. Detached from its only experimental base his behavioral plan mutates quickly into a dream of unrestricted personal authority to dictate social order. Having staked his survival on the supremacy of the will and the denial of 'extrahuman anxieties,' the unprogrammed loss of Slothrop triggers a return of the repressed and his attendant retreat into the 'solitude of a *Führer*.' Slipping 'into a *l'état c'est moi* frame of mind,' Pointsman hallucinates a fantasy of unlimited egoism from the vantage of a god: 'who *else* is doing anything? *isn't* he holding it all together, often with nothing beyond his own raw will ...?' When Dennis Joint asks 'hasn't he gone a bit mental suddenly?,' Pynchon explains: 'He has lost control' (*GR* 144, 227-9, 272-7).

After this, Pointsman is rarely seen again, his dreams of *personal* omnipotence an embarrassing and inefficient remnant. He makes occasional efforts to raise more funds over 'alcoholic luncheons with his various industrialists' but his usefulness is at an end. With the Anglo-American recovery of the Peenemünde rocket archives and the Nordhausen *Mittelwerk* hardware, Slothrop loose in the Zone no longer worries Pointsman's backers at Shell–Mex and ICI. It is Pointsman himself who proves a liability

---

45. PISCES stands for Psychological Intelligence Schemes for Expediting Surrender – 'whose surrender is not made clear' – and is a 'catchall agency' within 'The White Visitation' (34).

when his vengeful yet consistent plan to castrate Slothrop is bungled and American Major Duane Marvy is sterilized by mistake. While the castration of Marvy expresses the logical end of Pointsman's program – the engineered utopia requiring no armies or police – it puts him in disgrace with Clive Mossmoon and Sir Marcus Scammony of ICI.[46] His dreams do not square with the practical problems of international capitalism which needs to keep traditional authority in reserve. As Sir Marcus puts it, 'Slothrop was a good try at a moderate solution, but in the end it's always the Army, isn't it?' (*GR* 533, 615).

Pointsman's job is done and he, like his patient, is frozen in time. Where Tyrone Slothrop accumulates identities only to be dismantled after Potsdam, Pointsman develops experimental power only to yield it to higher authority. Just as Slothrop fragments into the seeds of post-war personnel and comes to rest as a powerless fool, so Pointsman is discarded and his discoveries seized for future use by that alliance of business and state power which emerges victorious in 1945. He is last seen sitting alone in a small PISCES 'sub-ministry' office provided by the authorities 'more out of sympathy than anything else,' nostalgic for his dogs and for the Nobel Prize he will never receive. Pointsman is, in effect, himself pensioned off like a faithful old dog (the appropriate product of a control system he helped advance), a powerless would-be *Führer* who terminates in common impotence with his preterite victim Tyrone Slothrop. Meanwhile, Sir Marcus Scammony takes up his Pavlovian program for application to the coming problems of the decolonization of the British Empire – '*our* lovely black animals' – and leaves Pointsman surrounded by 'an agency studying options for nationalizing coal and steel' in Britain. Capitalism is already leaving the war behind, accommodating its lessons, achievements, and requirements in order to routinize its operations in the interests of its long-term survival (*GR* 615-16, 752).

Pynchon's analysis of the dialectics of freedom and control within the development of capitalism moves towards a partial synthesis at the end of the third and pivotal section of *Gravity's Rainbow*. While the anarchic cores of both Slothrop and Points-

---

46. In Skinner's *Walden Two* (p. 149) Frazier tells Castle 'We don't *use* force! All we need is adequate behavioral engineering.' The necessary correction is provided by Dr. Benway in William Burroughs' *The Naked Lunch* (p. 54): 'A *functioning* police state needs no police.'

man are discarded and frozen in time, their usable residues are condensed out and reassembled in a single flexible container of the future. Through the figure of plastics man Clive Mossmoon, Pynchon draws together the scattered shells of Tyrone Slothrop and the behavioral program of Edward Pointsman: the differentiated personnel of post-war society and the key to their administration. Where the time limits of World War II and the blunders of the Secret Service once exploded the link between professor and patient, boss and worker, *Führer* and preterite victim, now the emergence of Mossmoon dramatizes the fusion of these dualisms in an engineered hybrid which rejects their destabilizing, irreconcilable aspects. Clive Mossmoon embodies the next moment in the unfolding of advanced capitalism's logic, further dissolving the spheres of private autonomy and public *praxis* into that of social behavior, and, by shifting the locus of control from outside to inside, transforming history into nature.[47]

Sitting in their club, the experienced Sir Marcus Scammony calms Mossmoon's anxieties about the election of the new Labour government, growing unrest amongst the Imperial Preference lobby inside Fleet Street, the Federation of British Industries, and the Conservative Party, and the attitude of the Americans towards 'socialism' in Britain. He introduces his protégé to the inner reaches of the international power structure – the Cliveden set being Pynchon's implicit historical instance – which has not only survived the war but which has thrived on its necessities and which operates on the principle encapsulated in Ishmael Reed's dictum: 'Politics are for poor people, the rich back all sides.'[48] Through the reassured Clive Mossmoon, Pynchon maps out some of the crucial intersections of the social system that emerges from the receding tides of World War II:

> Clive Mossmoon feels himself rising, as from a bog of trivial frustrations, political fears, money problems: delivered onto the sober

---

47. See the remarks of the 'spirit', Roland Feldspath, on the question of the interiorization of command in *Gravity's Rainbow*, p. 30.

48. Concerns about the abandonment of the Ottawa system, which united such strange bedfellows as Beaverbrook Newspapers, the Labour left, and the farm lobby, are detailed in Richard N. Gardner, *Sterling-Dollar Diplomacy* (New York: McGraw-Hill, 1969), ch. 3. On the Cliveden set, see Shoup and Minter, pp. 11-50 and Kees van der Pijl, *The Making of an Atlantic Ruling Class* (London: Verso, 1984), pp. 38-40, 110-15. Ishmael Reed, 'From The Final Appeal,' *CoEvolution Quarterly* 19 (1978), p. 67.

shore of the Operation, where all is firm underfoot, where the self
is a petty indulgent animal that once cried in its mired darkness.
But here there is no whining, here inside the Operation. There is
no lower self. The issues are too momentous for the lower self to
interfere ... No joy, no real surrender. Only the demands of the
Operation. Each of us has his place, and the tenants come and go,
but the places remain. (*GR* 616)

Mossmoon, the synthesized behavioral unit of post-war
society, routinizes the new order by containing the contradictions
of its emergence within a structure of rational adjustment.[49] With
most of the V−2 documentation retrieved from Dörnten and the
hardware being tested at Cuxhaven, the failure to recover Slothrop loses significance; with the maintenance of US control over
the British economy, the threat of radical socialism dissolves; and
with the election of a Labour government containing Attlee,
Bevin, and Dalton therefore committed to stabilize and polish the
capitalist system through public finance, bureaucracy, and
planning, the cash shortages which plagued Pointsman become,
if not a thing of the past, then no more than a relatively short-
term collar on the future. Technical knowledge, political influence, and economic power are drawn together to serve the needs
of the incipient totalitarian Operation. Inside, Mossmoon constitutes its personnel factor: the standardized agent of a standardized system of ostensibly peaceful production, consumption,
and exchange first anticipated in the plants and planning of
Henry Ford from 1913 onwards and later reproduced across the
world as the dynamic pole of a global 'Happyville.'[50] Apparently
stripped of all anarchic dreams of personal dominion, strength, or
suicide; secured within by an armored character immune to
alcoholic self-pity or irresponsible foolishness; and neutered by
labor and controlled consumption after years of more formal
rationing, Mossmoon is the integrated circuit at the end of
capitalism's own history: a temporary lodger in commercial
accommodation; an interchangeable part inside a technical
system; a morphologically differentiated cell adapted to a
naturalized environment. He embodies what Slothrop, the lazy
villager, could never be. For whilst Slothrop, through his senti-

---

49. Cf. Mailer, *Presidential Papers*, pp. 51-2.

50. See Antonio Gramsci, 'Americanism and Fordism,' in *Selections from the
Prison Notebooks*, trans. and ed. Quintin Hoare and Geoffrey Nowell Smith
(London: Lawrence & Wishart, 1971), pp. 298-313; van der Pijl, pp. 18-22, 90-91.

ments of love and preterition, retains something of his pre-totalitarian identity and is therefore dismembered and replaced, Mossmoon emerges as the City's productively careering executive, fully prepared for his role in 'this latest War' where any left-over foot-soldier friendship is dissolved and where 'the real and only fucking is done on paper' (*GR* 616).

Yet even while Clive Mossmoon rises from the receding tides of World War II onto 'the sober shore of the Operation' as an apparently stable, adjustable, and well-governed behavioral unit, the walling up of his 'mired darkness' indicates a continuing process of repression behind the Operation's slick corridors. 'Clivey' may contain the contradictions exposed by the separation, isolation, and dissolution of Pointsman and Slothrop but, 'chained and corseted though he be,' his is no terminal synthesis. The roots of the continuing tensions are to be found well beyond the 'shackled walls' of Sir Marcus's 'chastisement room' at 'The Birches.' They reach back into the Zone of occupied Germany. For as incipient totalitarianism's personnel factor, Clive Mossmoon inherits and restructures characteristics from all sides engaged in the transitional struggle ending in 1945, not simply from the Allies. Tyrone Slothrop is not the only object to break up on re-entry, to decay into or disperse across the shores of post-war. Nor is his connection with Pointsman the sole bond to be severed in the rationalization of the power ritual revealed by the war's end. Nor still are Slothrop and Pointsman the only ones to be sacrificed and passed over in the absorption of imperialism by incipient totalitarianism. Radicles of the Nazi formulation are also condensed into the new synthesis and consequently influence Mossmoon's characteristics (*GR* 616).

## Dominus Blicero: The Dismasted Man

> The basic problem ... has always been getting other people to die for you.
>
> — Thomas Pynchon, *Gravity's Rainbow* (1973)

The Axis phase of this transition reaches its critical point late in Pynchon's epic. In the final frames of *Gravity's Rainbow* a V–2 rocket, code number 00000, silently re-enters the post-war skies and accelerates down towards the roof of an old movie theater. Launched from Luneburg Heath in early May 1945 by an artillery battery under the command of Lieutenant Weissmann (whose SS code name is Dominus Blicero), it is like no other V–2 fired during the war. For the 00000 has been modified to include an insulation device – nicknamed the 'Schwarzgerät' – made of the mysterious plastic Imipolex-G. This synthetic fairing encases Blicero's young lover Gottfried, who has been 'mated' to the V–2 under the engineering eye of his commanding officer just before the countdown. The launching of the 00000 explodes the bond between Blicero and Gottfried on the Axis side just as the blunders of the Secret Service sever the connection between Pointsman and Slothrop on the Allied side. Indeed, the firing constitutes the ultimate moment in a German power relationship which precisely parallels the Anglo-American link not only in structure but also in development and dissolution. Gottfried, like Tyrone Slothrop, is a partly willing and partly unwitting plaything of power who is finally dispersed through a Christian sacrificial ritual, the youthful and innocent victim required for the survival of any control system whose necessity and pleasure lie in its own proliferation. Blicero, like Pointsman, is an agent of power whose authoritarian dream is at once revealed and dismembered by the loss of his child victim; a would-be *Führer* who is himself rendered impotent in the interests of the Operation's survival (*GR* 272, 322, 431-2, 625, 721-4, 749-51, 760).

But within this corresponding drama there exists a structural and historical asymmetry. For while, in the formal, nationalized context of World War II, Blicero and Gottfried represent the Axis in the same way that Pointsman and Slothrop represent the Allies, considered as agents of the larger transitional struggle they embody the last stand of the imperial order fighting a losing battle against the forces of incipient totalitarianism. This disparity is expressed in the personal and historical trajectories of Blicero and Pointsman which, even though they share common appearances and termini, are not identical. Pointsman is a more transitional figure than Blicero: he rises against and displaces the classic British imperialist Brigadier Pudding, advances the cause of civilian expertise against that of military nostalgia, and develops the rudimentary behavioral solvents of power which are taken up by the international business network after the suspension of military operations. Blicero, by contrast, is a child of imperialism who, while accommodating technical innovation, retains the imperial form and advances it to what is at once its ultimate expression, its terminus, and its transubstantiation. As such he shares its fate. For what is really defeated in 1945 is not so much the Axis as a specific historical power complex – imperialism – which is shown to be not only no longer sufficient for but on its own actually a threat to the long-term survival of the Operation.[51] With the capitulation of the Nazis recorded in the launching of the 00000, the stewardship of that Operation passes to the combined emerging forces of incipient totalitarianism which take as the spoils of victory the right to determine the administration of post-war civil society: the behavioral routine of Pointsman's 'Happyville.' By contrast, as the embodiment and apotheosis of superseded imperial power, Blicero suffers exile to 'Pain City' which, with its martial roots, is displaced to the psychological and spatial peripheries of the new civilian domain: the internalized private reaches of the armored self, the censored outhouses of the remaining colonies, and the distant edges of the skies (*GR* 317, 644-5).

In this sense he comes to rest in a position congruent to the one from which he began his trajectory. For just as he leaves the temporal field of *Gravity's Rainbow* at the trailing edge of the new incipient totalitarian order in 1945, the 'new cycle' which will find its 'Deathkingdom' on the moon, so Blicero enters it at the leading edge of the old imperial order whose killing ground was in southwest Africa some forty years before. More specifically, he enters the novel just as the events which foster the ultimate Nazi expression of that order – acceptance of the Versailles Treaty, the setting of reparations, the loss of Upper Silesia, inflation, and the occupation of the Ruhr – are starting to come together. Between these two congruent points Blicero's trajectory then registers and represents the dynamics of this last stand of imperialism (*GR* 723).[52]

Blicero first appears as Army Lieutenant Weissmann (or 'white man'), an agent of the imperial German fatherland – 'a very young man in love with empire' – who is despatched in 1922 on 'his own African conquest' to the colony of Südwest. There he seduces a Herero boy and christens him 'Enzian' before bringing him back in 1926 to metropolitan Germany as 'his own faithful native' and 'pet.' Nurtured by the manly pursuits of secret rearmament, this domesticated colonial bond grows deeper with

---

51. See Gramsci, pp. 293-4; van der Pijl, pp. 1-2, 106.
52. See Alan Bullock, *Hitler: A Study in Tyranny*, rev. edn (Harmondsworth: Penguin, 1962), pp. 85-91.

the accession of the National Socialists – whose appeal to folkish tribal unity coupled with 'devotion to the leader' hold Enzian's allegiance as much as they satisfy Blicero's authoritarian desires – and thrives on the early military conquests leading to World War II. But with the rise of Nazism signalling not simply the apotheosis of imperialism but through it – pupating inside the imperial chrysalis – the emergence of a new historical formation, that bond becomes increasingly strained. Having accepted Blicero as the god-like embodiment of simple and innocent love in Südwest, Enzian discovers in the northern hemisphere that his dream of tribal unity is only one moment within the imperial complex and that now, with early successes giving way to military defeats on all fronts, it is part of a synthesis approaching dissolution. So whilst he remains satisfied by 'the simple feel and orgasming' of his love affair with the white man ('it was still very physical for me then,' he later recalls), he finds that Blicero 'had already moved past that part of it' and on to the routinized and fetishized *Gesellschaft* of the totalitarian transition, where 'love ... had to do with masculine technologies, with contracts, with winning and losing,' not with the simple coupling and sharing – 'all sets of opposites brought together' – of the divine *Ndjambi Karunga* in the Südwest (*GR* 99-100, 322-4, 352, 360).[53]

To sustain Blicero's love Enzian is obliged to 'enter the service of the Rocket' which, as the harbinger of incipient totalitarianism, has begun to displace their colonial bond. But for this 'first step towards citizenship in the Zone' he is racially, psychologically, and culturally unsuited. For beyond 'simple steel erection, the Rocket was an entire system *won*, away from the feminine darkness, held against the entropies of lovable but scatterbrained Mother Nature' (*GR* 324). Enzian, by contrast, remains a part of his own Herero past, with its covalent *Gemeinschaft* restraining the lure of Blicero's ionic accumulation. Finding himself unable to 'see into him or the things he believed in,' the Herero becomes increasingly estranged from Blicero as the turning tides of the war accelerate the erosion of the imperial power structure: 'As the Rocket grows toward its working shape and fullness, so does he evolve, himself, into a new configuration.' Thus in February

---

53. The Herero term is translated variously as the name of God and the attainment of heaven or origin. Blicero's movement is reflected in his adaptation of an SS code name to replace the name that suited him in the Army. See Arendt, *Origins*, pp. 380, 419-37.

1945, during the evacuation of their rocket team from Peenemünde to the *Mittelwerk* at Nordhausen, Enzian drops out of
sight altogether to go underground and lead the Schwarzkommando, those Hereros who were brought to Germany mostly
after 1933 to be trained as rocket experts and later used to administer post-war colonial 'shadow-states' for the Nazis, and who are
now exiles hiding in the Zone's abandoned mineshafts. Their
dream, under the Moses-like guidance of Oberst Enzian, is to act
as 'scholar–magicians of the Zone' who, in the course of assembling their own rocket (code number 00001), will accrue 'information and expertise' sufficient to help them replace 'absolute
weaponry' with communal salvation (*GR* 315, 318, 324-5, 427,
520).[54]

Blicero, meanwhile, begins to register the accelerating
fragmentation of the imperial order. Faced with the continuing
encroachment of Allied forces he falls back deeper into the service
of the *Vergeltungswaffen* which both he and his superiors hope will
turn the tide and preserve their old order. During October 1944,
he starts supervising an artillery battalion launching V–2s from
The Hague against Antwerp.[55] As a result of this move Blicero
also finds a way of stiffening his personal imperial bonds.
Deprived more and more of his original native boy, he replaces
Enzian with a young and innocent Army conscript called
Gottfried who is serving with the Hague artillery battalion. Docile
and obedient like the young Herero, Gottfried enjoys his
commanding officer's discipline, punishment, and fetishism, and
so becomes the new 'young pet and protégé of Captain Blicero,'
Enzian's northern hemisphere equivalent. And yet the familiar
form of this relationship masks its changing content, for Gottfried
is seduced into captivity not, like Enzian, by Blicero himself, but
by the rockets – which 'are his pet animals' – and by the thrills of
war they express. Moreover, unlike Enzian, Gottfried is white'
and German. Under the Nazi transition the imperial bond is

---

54. Enzian's separation from Blicero is indicated by his name. The *Enzian* was
a German surface–to–air missile designed in 1943 but cancelled in February 1945
because of its poor experimental performances and the need to concentrate
resources on the V–1 and V–2 programs. Both Enzians were thus passed over
for the *Vergeltungswaffen*. See Frederick I. Ordway III and Mitchell R. Sharpe, *The
Rocket Team* (London: Heinemann, 1979), p. 80; Brian Ford, *German Secret
Weapons: Blueprint for Mars* (1969; London: Pan, 1972), pp. 131, 148-51.

55. Details of actual V–2 battery deployments are given in Ordway and
Sharpe, pp. 203-5.

taking on the appearance of a shell protecting a new incipient totalitarian embryo, or of the albumen which, as the ovum grows and hatches, is absorbed and hence survives within the resultant fledgling. Blicero's struggle for survival, which expresses the last stand in imperial society's defense of its private and public structure, is therefore being undermined from within as well as without (*GR* 94-103, 484).

During the spring of 1945 this struggle enters its final phase. Dramatized through the trinity of Blicero, Gottfried, and the 00000, it has two interrelated but polarizing plots which record, on the one hand, the fragmentation of the old order's private and public dimensions under the emasculating pressures of their own contradictions, and on the other, the process of ejection and assimilation they experience in the baptismal fires of the new. The first plot is Blicero's own; the second has as its subject the sacrificial deliverance of power from Blicero into the new synthesis of Gottfried, Imipolex-G, and rocket. The two cross and then rush apart at the launching of the 00000 on Luneburg Heath. The site is the clearing where imperialism and incipient totalitarianism commune, where the Father erects, mounts, and offers up his Son, and where the Operation delivers its real patient – the 'virus of Death' – from the evils of historical decay.

Blicero's plot is a monologue of withdrawal. The period between late March and early May – 'a time of dissolution,' Miklos Thanatz later recalls – sees 'things . . . falling apart' across the Zone. Blicero, who as the vehicle of imperialism had struggled to hold them together in the face of internal and external retraction, shares in this dissolution by finally abandoning the public sphere to the advancing Allied forces and falling back into the isolated and 'clotted islands' (or 'I-Lands') of his privatized colonial dream world. Unable to maintain that social intersection which holds public history and private autobiography in a dialectical process, he tries to escape its historical transformation by dreaming of an end to history, 'of discovering the edge of the World. Finding that there *is* an end.' He reacts to the dissolution of the imperial order, that is, by inaugurating himself as some ahistorical terminus, thus cancelling its unresolvable contradictions by dissolving the flux of history into the permanence of autobiography: projecting a personal world out of his own fantasized nature and thus denying his own historical displacement (*GR* 465, 485, 722).

Pynchon dramatizes this withdrawal, in the recollections of two of its witnesses, through a set of interlocking metaphors and images – mythological, musical, zoological, cinematic – which articulate the political and psychological drives Blicero embodies at the heart of the imperial order. Sailing down the Oder river on board the *Anubis* in early August 1945, Miklos Thanatz recounts to Slothrop how, as they fell back towards Luneburg Heath some months earlier, Blicero became a 'screaming maniac':

> He'd begun to talk the way the captain in *Wozzeck* sings, his voice breaking suddenly up into the higher registers of hysteria ... and he reverted to some ancestral vision of himself, screamed at the sky, sat hours in a rigid trance, with his eyes rolled clear up into his head. Breaking without warning into that ungodly coloratura. White blank ovals, the eyes of a statue, with the gray rain behind them. He had left 1945, wired his nerves back into the pre-Christian earth we fled across, into the Urstoff of the primitive German, God's poorest and most panicked creature. (*GR* 465)

The following day Greta Erdmann tells Tyrone how at Luneburg Heath in late April or early May she and Thanatz found Blicero 'in his final madness.' By then, she recalls, he had 'grown on, into another animal ... a werewolf ... but with no humanity left in its eyes.' These 'wrinkled wolf-eyes had gone ... on into its animal north, to a persistence on the hard edge of death ... tough cells with the smallest possible flicker inside.' Shrieking quietly, Blicero had told her that they reflected 'the map of my Ur-Heimat, the Kingdom of Lord Blicero. A White land.' And in turn Greta had come to her own 'sudden understanding' of Blicero's condition: 'he was seeing the world now in *mythical regions*: they had their maps, real mountains, rivers, and colors. It was not Germany he moved through. It was his own space' (*GR* 486).

The 'last stand in the Luneburg Heath' is approached as an attempt to slow the flicker of imperialism's social movie and to bring its temporal decay to a standstill in a personally frozen frame filled by the map of Blicero's own Kingdom of time and space. Within his dream, Blicero takes on a 'mineral conscious-ness' moving at 'frames per century, per millennium,' and petrifies into the rigid and domineering type at the center of the imperial apotheosis. Exiled from the public sphere, losing even his edited shots of Miklos Thanatz and Greta Erdmann, Blicero retreats to the mythological ancestral source of the Ur-Heimat to plunge his climactic identity into the primal material of the Urstoff – the impulse towards collective extinction which feeds the Nazi state – and thus to retrieve the tribal depolarization submerged by the onset of the new order's mediated and contractual routines.[56] As he approaches this wellspring, so he mutates through other mythological forms – from 'local deity' to 'werewolf' – until he appears as an invisible and terrifying minotaur who monitors the shores of his remaining imperial island–labyrinth before turning to confront the advancing forces of incipient totalitarianism (led by Pointsman's dream Theseus) in a struggle for control of the sacrificial system centered on the 00000 and Gottfried (*GR* 142-3, 465, 485-6, 612, 666, 721).

With this first plot approaching terminal dispersion, the second comes into focus, gains momentum, and demonstrates how in attacking the myth of origin as an essential element of fascist ideology, Pynchon's text reveals the Ur-Heimat to be not 'his own' but a *socially produced* space. In late March, as his artillery battalion is forced to pull back eastwards to avoid capture, Blicero travels to the *Mittelwerk* and directs rocket engineer Franz Pökler (whom he had transferred there from Peenemünde the previous year) to build in complete secrecy a modification – the Imipolex-G fairing – for the propulsion section of his last V–2, the 00000. He has decided on a final act, the firing of Gottfried, which will not only consummate and discharge his remaining imperial bond but also ensure its deliverance into the new postwar order (*GR* 143, 427, 431, 721).

As the elaborate launching preparations approach, Blicero elegiacly addresses the innocent Gottfried on the dream of immortal and transcendental purity which will require his total submission:

> Can you feel in your body how strongly I have infected you with my dying? I was meant to: when a certain time has come, I think that we are all meant to. Fathers are carriers of the virus of Death, and sons are the infected ... and, so that the infection may be more certain, Death in its ingenuity has contrived to make the father and son beautiful to each other as Life has made male and female ... oh Gottfried of course yes you are beautiful to me but I'm dying ... I want to get through it as honestly as I can, and your immortality rips at my heart – can't you see why I might want to destroy that, oh that *stupid clarity* in your eyes ... when I see you in morning and evening ranks, so open, so ready to take my sickness in and

56. See Mailer, *Presidential Papers*, pp. 147, 198-9, and the conclusions of scientific intelligence expert R.V. Jones on the appeal of the V–2, quoted in David Irving, *The Mare's Nest*, rev. edn (London: Panther, 1985), pp. 299-300, 327-8.

shelter it, shelter it inside your own little ignorant love. . . . I want
to break out – to leave this cycle of infection and death. I want to
be taken in love: so taken that you and I, and death, and life, will
be gathered, inseparable, into the radiance of what we would
become. (*GR* 723-4)

This dream of absolute power involves an absolute sacrifice in
order to deliver the 'virus of Death' (which will through characteristically twentieth-century images of mass physical invasion
spread Mailer's and Reich's post-war totalitarian plague) across
the boundaries of imperialism.[57] With 'American death . . . come
to occupy Europe,' the latter's 'impulse to empire, the mission to
propagate death, the structure of it,' has reached its 'last phase.'
It has been reduced to Blicero's private sacrificial oven. But a new
state and a new cycle to carry that virus to its next 'Deathkingdom' on the moon are ready to begin. For as Blicero delivers his
infected child into the 00000, so the imperial impulse is absorbed
into the incipient totalitarian order. The armbands, stormtroops,
death camps, and other manifestations of imperial society's last
stand, like the 'great rainbow plumes . . . fittings of gold [and] epic
marches over alkali seas' of its prime, are jettisoned as unnecessary and inefficient, to be replaced by the smooth, polished steel
and plastic of the Imipolex-cushioned Rocket State. But inside
the latter's parabola, the seed of a straitjacketed *Führer* persists.
As it extinguishes Blicero's 'Little State,' the firing of Gottfried in
the 'black indomitable Oven' of the 00000 also constitutes the
'birth scream' of incipient totalitarianism. The delivery of
Gottfried from the Imipolex womb into the skies and across the
post-war world signifies the survival and dispersal of the 'virus of
Death' from its previous incarnation (*GR* 99, 722-4).

The 'civil paradox' of Blicero's 'Little State,' that its 'base is the
same Oven which must destroy it,' is exploded and resolved by
the new synthesis, and its minotaur–commander monumental-
ized and passed over. He is found, after the launching, by Miklos
Thanatz who, having witnessed the sacrifice and its aftermath in
May, travels with dramatic implausibility by 'purloined P–51
Mustang,' the *Anubis,* and various Displaced Persons trains –
'round and round the occupation circuit' – until he is discovered
by Enzian's Schwarzkommando in late August (*GR* 99, 663-71).

---

57. See Mailer, *Presidential Papers*, pp. 198-202; Reich, *Character-Analysis*, pp.
248-80.

In transit, as he recovers from the dissolution of the Oven State he had surrendered to, he starts to recall what he found beyond Blicero's zero:

> After the last firing ... Thanatz came to imagine he had disposed of Blicero ... in that ... very conditional, metallic way. And sure enough, the metal had given way to flesh, and sweat, and long chattering night encounters, Blicero cross-legged stammering down at his crotch I cuh-cuh-cuh-cuh – 'Can't,' Blicero? 'Couldn't'? 'Care'? 'Cry'? Blicero that night was offering all his weapons, laying down all maps of his revetments and labyrinths. (*GR* 672)

Blicero has lost the power of his colonial bonding. He has surrendered his Gottfried, God's peace, piece, or penis, to the new order and is left, as Enzian interprets his last letters, a 'kind of cripple'; like Ahab in Melville's *Moby Dick*, 'a dismasted man' who 'never entirely loses the feeling of his old spar.' Degenitalized and deprived of his devilish powers, Blicero loses his petrified stiffness along with his hooves and horn, his protective armor, and the other scales of his mythical kingdom. Like Pointsman on the Allied side, having served the interests of the Operation, his claims of *personal* omnipotence render him of no further use and he is abandoned as an outmoded relic. With the rocket arching into the post-war skies, Blicero is left to regain his 'human consciousness, that poor cripple, that deformed and doomed thing,' and to rest in misery while his joyless offspring heralds 'an eternally progressive progeny of griefs' beyond the Oven State's grave.[58]

## Clive Mossmoon: On the Beach

> Yes, giant rubber cocks are here to stay as part of the arsenal.
>
> — Thomas Pynchon, *Gravity's Rainbow* (1973)

The nature and potential destination of Blicero's offspring adumbrates the structure and domain of the Operation's new synthesis: the Rocket State of incipient totalitarianism. Its nature

---

58. Melville, *Moby Dick*, pp. 575, 582. See Bush, pp. 51-7 for the analysis of Ahab drawn on here.

is clear in Enzian's remarks to Katje Borgesius in late August 1945, based on Blicero's last letters to the Herero, which summarize and project the processes of sacrifice and deliverance in this malignant Passover:

> My slender white adventurer, grown twenty years sick and old ... was changing, toad to prince, prince to fabulous monster ... 'If he is alive,' he may have changed now past our recognition. We could have driven under him in the sky today and never seen. Whatever happened at the end, he has transcended. Even if he's only dead. He's gone beyond *his* pain, *his* sin – driven deep into Their province, into control, synthesis and control ... (*GR* 660-61)

Blicero, his personal trajectory complete and his energies exhausted, has yielded up power through his own extermination; the 00000 now passes overhead as the latest vehicle of the new synthesis, the latest redemption for the Operation and its agents. As for its potential destination, the coming territory of the rocket is as pervasive and extensive as the promises implicit in the answers Blicero never has to give to his youngest child before the sacrifice in the clearing. For the point at which Gottfried should seek the meaning of the Passover, the 'wind-beat moment' with 'the iron rockets waiting outside' in which his dissent might foster resistance elsewhere, is given as the archetypal instance of personal responsibility which can appear at any point within the post-war order: 'The scene itself must be read as a card: what is to come. Whatever has happened since to the figures in it ... it is preserved, though it has no name, and, like the Fool, no agreed assignment in the deck.' As a result of the casting of Blicero's tarot before the launch, this unspecified card is given a name – 'The World' – and thus a domain in which to exercise its indeterminate range. Hence, unlike the restricted Oven State of the late imperial era, the Rocket State becomes pervasive. Synthesizing itself around the 00000 which, in the final frames of *Gravity's Rainbow*, accelerates down towards the roof of the Orpheus Theater in contemporary Los Angeles, it spends more than a quarter of a century passing over and integrating the industrialized and peripheral world into its domain (*GR* 724, 746-9).

But if the Rocket State polymerizes around the 00000 it must first synthesize its major components – the basic V–2, Gottfried, and the Imipolex-G shroud, all of which have previously been produced within and liberated by the fragmentation of the Oven State – into an effective compound. The combination of Gottfried and the rocket by means of the flexible Imipolex sheath embodies this synthesis. By fusing the autobiographical seed of Blicero with the technical shell of the rocket under the adjustable stimulus of Imipolex-G, it reintegrates private and public moments of power in a new social nexus contained and directed by a new agent of integration. As a result, it reinserts those dreams of victory and submission latent in any society back into an historical vehicle able not only to contain but also to reproduce and breed them. It therefore provides the ideal agent for the new order's expansion; in Mailer's terminology, it constitutes the viral carrier of the totalitarian plague.

In a sense, Imipolex replaces Blicero as the routinely manufactured, non-human bonding agent of incipient totalitarian society. For unlike Blicero, who turned as rigid and static as a dead man or stiff, Imipolex 'is the first plastic that is actually *erectile*,' a protean 'Peculiar Polymer' that under suitable stimuli can adapt itself to any form 'from limp rubbery amorphous to amazing perfect tesselation, hardness, brilliant transparency, high resistance to temperature, weather, vacuum, shock of any kind.' It is the ideal material for a Rocket State which can 'orchestrate sudden discontinuities – blows, wrenchings – in among [its] more caressive moments'; for a welfare–warfare state able to move quickly between 'repressive tolerance' and punishing intolerance, between its official 'Happyville' and its persisting inner 'Pain City'. From nurturing the seeds of this 'latest spring' inside the 00000, Imipolex will therefore nourish and sustain the new cycle's later shoots until their flowering on the moon (*GR* 699).

With the firing of the 00000 the second plot in Blicero's – and hence the imperial order's – abortive struggle for survival is almost complete. Blicero has been passed over and his offspring is experiencing a rapid upbringing after the labor and delivery in the clearing. Yet the proceedings are not quite over. For beyond the *Brennschluss*, or rocket motor cut-off, Blicero's seed still has to re-enter the post-war order; his 'virus of Death' still has to find its new host. The re-entry of the 00000 marks the final phase of this transition. By incorporating imperialism's residual burnt offering within the incipient totalitarian body it draws together and reconstitutes all those liberated radicles, shells, nuclei, and solvents of society which survive the war's end. This process is registered on both sides of the dissolving historical divide, in the 00000 and within its new domain, as well as on both sides – public and private – of the dissolving structural divide. The combination of propellants in the V–2, which is only made possible by the pre-war international corporate cartelization movement (the harbinger of the Operation's post-war incarnation),[59] indicates the reintegration of public power within the Rocket State by projecting a technically synthesized control structure accommodating and governing all so-called ideological conflicts: 'the Rocket ... is able to gather violent political opposites together ... as it gathers fuel and oxidizer in its thrust chamber: metered, helmsmanlike, for the sake of its scheduled parabola.' At the same time the resulting motion of the rocket expresses the private dimensions of this synthesis. For just as the separation, dispossession, and fragmentation of Pointsman and Slothrop yield advanced capitalism's ideal behavioral prototype in the shape of Clive Mossmoon, so the re-entry of the 00000 expresses the liberation and deliverance of its 'lower self' – an ironic 'return of the repressed' – and thereby completes his personal integration (*GR* 318).[60]

In Gottfried's sacrifice Pynchon activates a synthetic arc under which public history and personal identity may find renewed social coherence. But it is a coherence whose precarious structure is only emphasized by the precipitation of Clive Mossmoon. For Mossmoon is a complex figure of post-war containment in whom incipient totalitarianism's dream of terminal stability interacts dialectically with its anarchic yield; a mass man whose work for the Operation co-exists with his 'leisure' activities in the 'chastisement room' at Sir Marcus Scammony's estate 'The Birches.' Moreover, as an agent of that Operation, Mossmoon is being prepared by Sir Marcus to foster its – and hence his own – extension. As he and the Rocket State polymerize across the post-war world, so they spread a civil society in which both public and private power constitute one another only in a radically unstable symbiosis; an advanced capitalism whose trajectory matches that of the rocket itself: permanently decaying, approaching and avoiding, falling ever closer to the last delta–t (*GR* 616).[61]

---

59. On the relationships between the international cartel network, fuel production, and German military operations, see Joseph Borkin and Charles A. Welsh, *Germany's Master Plan: The Story of Industrial Offensive* (London: John Long, 1943), pp. 85-90; Joseph Borkin, *The Crime and Punishment of I.G. Farben* (New York: Macmillan/The Free Press, 1978), pp. 46-94; Gabriel Kolko, 'American Business and Germany, 1930-1941,' *Western Political Quarterly* 15 (1962), pp. 718-27; Charles Higham, *Trading With the Enemy* (London: Robert Hale, 1983), pp. 32-62.

60. Cf. Norman Mailer, *Pieces and Pontifications* (1982; Sevenoaks: New English Library, 1983), Part 1, p. 157.

61. The precise course of the re-entry of Blicero's seed into Clive Mossmoon is discussed in Appendix 2.

## Synthesis and Control

> And there's that *smell* again, a smell from before his conscious
> memory begins, a soft and chemical smell, threatening, haunting,
> not a smell to be found out in the world – *it is the breath of the
> Forbidden Wing.*
>
> — Thomas Pynchon, *Gravity's Rainbow* (1973)

Mossmoon, like his collective ancestry, remains no more than an
instrument of the power structure, the agent of an imagined
Operation whose priorities extend across both individual
lifetimes and national boundaries. The historical roots of this
Operation reach back beyond the covers of Pynchon's epic into
the salons of the European imperial elite and, more specifically,
to the World War I work of Walter Rathenau, head of AEG, the
huge German electric power and equipment combine, and director of over 100 large European corporations. Rathenau, who
combined his business interests with a career in the academic
world and in politics (he later became German Foreign Minister),
was 'prophet and architect of the cartelized state.' At the
outbreak of war, with the German General Staff committed to the
outdated imperial Schlieffen Plan of swift victory against France
and then Russia, Rathenau alerted Minister of War Erich von
Falkenhayn to the desperate need for industrial mobilization and
– in the face of the British naval blockade – for raw materials
security against a prolonged conflict. His comprehensive proposals led to Rathenau's appointment as head of the War Raw
Materials Office within the Ministry of War, and a consequent
coordinating role over the entire German war economy. In Berlin
he devoted himself, in Pynchon's words, to 'controlling supplies,
quotas and prices, cutting across and demolishing the barriers
that separated firm from firm' in the interests of victory. Rathenau was 'a corporate Bismarck,' and although the British blockade finally forced the German capitulation (as he had foreseen) in
1918, his influence, like Bismarck's, persisted beyond nations and
wars (*GR* 164-5).[62]

In *Gravity's Rainbow* Rathenau is presented as the first influen-
tial example of the Operational mind:

---

62. Borkin, *Crime and Punishment*, pp. 10-11, 27.

Young Walter was more than another industrial heir – he was a philosopher with a vision of the postwar State. He saw the war in progress as a world revolution, out of which would rise neither Red communism nor an unhindered Right, but a rational structure in which business would be the true, the rightful authority – a structure based, not surprisingly, on the one he'd engineered in Germany for fighting the World War. (*GR* 165)

According to Pynchon, Rathenau's vision of an apolitical global business system motivated his post-war political career and his conclusion, as German Foreign Minister, of the Rapallo Treaty with the Soviet Union in 1922 (*GR* 166, 338). The Rapallo Treaty linked Bolshevik Russia with Weimar Germany in a marriage of convenience designed to undermine the effects of the western commercial boycott and diplomatic isolation of the USSR and to evade the restrictions placed on German industrial and military power by the Treaty of Versailles. It included agreements on diplomatic recognition, financial relations, trade and investment, and (unofficial) military collaboration between the two countries, and as such constituted one of the first concrete signs of the *Interessen Gemeineschaft*, the community of interest, between capitalist and communist powers that Rathenau had mapped out.[63]

The German Foreign Minister's 'opposite' number in the treaty negotiations was George Chicherin, the Soviet People's Commissar for Foreign Affairs. A 'long-term operator', Chicherin constitutes a second important figure in Pynchon's history of the Operation's expansion: he provides a blueprint for the administrative superstructure of Rathenau's incorporated world into which Mossmoon's molded masses will slot. In Pynchon's words, Chicherin 'believed in a State that would outlive them all, where someone would come to sit in his seat at the table' and where 'sitters would come and go but the seats would remain.' He proposed a routinized world of natural administration; a system of permanent 'Control' – the 'fixed roulette wheel' Tyrone glimpses beneath the skirts of the 'Forbidden Wing' at the Casino Hermann Goering – which would absorb and neutralize political and ideological conflict in the interests of stability and operational continuity (*GR* 209, 338).

---

63. Adam B. Ulam, *Expansion and Coexistence: Soviet Foreign Policy, 1917-1973*, 2nd edn (New York: Praeger, 1974), pp. 146-52; Hans Gatzke, 'Russo-German Military Cooperation During the Weimar Republic,' *American Historical Review* 63 (1958), pp. 565-97.

But with the rise of Nazism in Germany this vision of a permanent state combining private and public authority in a single rational system found its immediate realization not in Europe but in the United States. Even as he negotiated the Rapallo Treaty with Chicherin, Rathenau, a liberal Jewish Republican, found himself the scapegoat of the German nationalists, blamed for the disgrace of Versailles – whose demands he and Chancellor Wirth had attempted to fulfil – for the huge reparations burden, and for the associated currency depreciation. Assassinated two months later, Rathenau was an early victim in the deepening struggle which brought Hitler (who had branded the treaty as 'Jewish Bolshevism') to power in 1933.[64] Nevertheless, his influence survived. In *Gravity's Rainbow* Rathenau advises the 'elite . . . from the corporate Nazi crowd' via séances on the structure of rational organization he sees immanent in the very nature of matter itself. To Generaldirektor Smaragd of I.G. Farben he poses two questions which confront the changing structure of power, questions which only the 'transcended' like Blicero can answer: 'First, what is the real nature of synthesis? And then: what is the real nature of control?' These questions reach directly into the productive center of *Gravity's Rainbow* since they concern modes of obtaining and deploying energy for the extension of political and economic power. However, they are answered not by the Nazi crowd – Heinz Rippenstoss, 'the irrepressible Nazi wag and gadabout,' can only interrupt to ask Rathenau 'is God really Jewish?' – but by an American businessman, Lyle Bland (*GR* 164-7).

Rippenstoss's characteristically Nazi preoccupation instances Pynchon's sense that the Third Reich of itself constituted a vital but highly volatile and ultimately dysfunctional catalyst for a larger reaction which, under the hothouse conditions of war, would precipitate a reconstituted Operation no longer dependent on the ideologies and terrors of the SS state. For whereas the United States between the wars introduced through mass production processes and the New Deal both the technological bases and politicoeconomic structures suited to the Operation's deliverance, Germany underwent a series of developments that would render it of no further use: from the imposition of punitive reparations and the removal of its colonial possessions at the

---

64. Bullock, p. 318.

Versailles Conference of 1919 to the extended financial and political crisis of the early 1930s; from the occupation of the Ruhr and
the hyperinflation of the early 1920s to Hitler's unilateral
program of *Endlösung* and war. For Pynchon, the establishment of
the Third Reich and its prosecution of the war certainly advanced
the Operation's long-term designs, imposing, for example, a *de
facto* integration of mainland Europe that would later be taken up
by American policy makers as a desirable post-war objective and
pressing the United States into an accelerated mass mobilization
of technology, men, and materials that would provide the necessary groundbase for the Operation's subsequent resurrection.
But in promoting these transformations the Nazi system itself
'remained confined to acting out a cruel caricature of a restructuration of the class structure by its genocide of the Jews and the
annihilation of working-class organizations.'[65] As Rathenau tells
his corporate Nazi contacts, while he had endeavored to further
the Operation's long-term needs, they were engaged in activities
that, notwithstanding their 'global implication[s],' were from the
Operation's point of view 'trivial side-trips.' Correspondingly, it
was not amid the wreckage of Europe that the late German
Foreign Minister's messages would be picked up and acted on
but in the long-preserved United States (*GR* 165-7).

In *Gravity's Rainbow* Lyle Bland is a Bostonian entrepreneur,
board member of the Slothrop Paper Company of Massachusetts, and Tyrone's uncle. He is involved in numerous international business transactions which connect him financially with
Hugo Stinnes, a real-life German financier and corporate empire
builder, and hence with Walter Rathenau. These transactions
range from a deal involving the Slothrop Paper Company in
providing private currency to Hugo Stinnes and 'mefo bills' to
Hjalmar Schacht's German Treasury (financing the secret
rearmament program) to another involving Stinnes, chemist
Laszlo Jamf, and the Grossli Chemical Corporation, which covers
the 'education' and long-term surveillance of Tyrone himself.
While Bland is a fictional character, Pynchon models his business
operations on the history of international corporate relations
between the wars, and in particular on the global rubber and
chemical cartel organized by I.G. Farben and the Standard Oil
Company from 1925, which not only assisted Hitler's rearma-

---

65. Van der Pijl, p. 106; Yergin, p. 473. See also Mumford, pp. 233, 250-52.

ment drive but also weathered and survived the 'unsettled conditions' of World War II. His extensive business interests lead Bland from playing all sides of the system into the Business Advisory Council, an historical organization established by Gerard Swope of General Electric in 1933 to advise President Roosevelt on the need for government cooperation in the rational, business-led planning of the US economy Swope believed essential to the recovery and long-term survival of capitalism. According to Arthur Ekirch, 'what Swope proposed was the planned cartelization of a large part of American industry, with certain guarantees and benefits to labor.' As Pynchon notes, these 'ideas on matters of "control" ran close to those of Walter Rathenau, of German GE.' Just as in the late nineteenth and early twentieth centuries the reformist welfare policies of Bismarck in Germany were adopted by governments in Britain and the United States, so during the next generation the programs of Walter Rathenau, the 'corporate Bismarck,' reappear elsewhere as common solutions to meet common necessities (*GR* 165, 284-5, 581).[66]

Lyle Bland's answers to Rathenau's questions from beyond the grave are never announced directly, but his proliferating schemes suggest that he has considered them carefully. For his commercial and financial maneuvers do not exhaust his energies. On the contrary, as Pynchon notes, the Bland Institution and the Bland Foundation have been pressing 'his meathooks into the American day-to-day since 1919' by providing philanthropic cover for his monopolization of technical advance, his corruption of the academic world, his manipulation of narcotics and advertising, his secret deals with the FBI, and his psychological abuse of human eroticism. In this sense Bland constitutes an historical summation or integral of the extension of business power across the social space during the twentieth century, all of which makes him a direct descendant of Walter Rathenau himself (*GR* 580-81).[67]

Characteristically, Bland's political support is given to the man whom unreconstructed laissez-faire capitalists like the du Pont family (themselves in partial competition with the I.G. Farben–

---

66. Arthur A. Ekirch, Jr., *Ideologies and Utopias* (1969; Chicago: Quadrangle, 1971), p. 53; van der Pijl, pp. 54-5, 69-70, 81-3, 88-9; Borkin, pp. 76-94. Standard Oil President Walter Teagle used the phrase 'unsettled conditions' to describe World War II (quoted in Borkin, p. 83).

Standard Oil cartel) and the Liberty League members consider a
fundamental threat to free enterprise: Franklin Delano Roosevelt.
Despite their opposition, for Lyle Bland,

> FDR was exactly the man: Harvard, beholden to all kinds of
> money old and new, commodity and retail, Harriman and
> Weinberg: an American synthesis which had never occurred
> before, and which opened the way to certain grand possibilities –
> all grouped under the term 'control,' which seemed to be a private
> code-word – more in line with the aspirations of Bland and others.
> (*GR* 581)

Roosevelt is a political *Verbindungsman*, or connection man, a
governmental copy of Wimpe at I.G. Farben. Linked to the vital
centers of the American business system, descended from an
established ruling-class family, well-versed in the ways of American government, and devoid of the anti-semitic ideology which
provided I.G. Farben with only short-term benefits and long-
term problems, Roosevelt is the ideal synthesis for delivering the
United States into the post-war world. His shortcomings in
economics and business sense allow the corporate leaders to plan
his education whilst his fantastic political skills ensure public
support for the privations and violence necessary to the
Operation's deliverance. He is the front man who, from the
depths of depression to the edge of victory, saves the day (*GR*
344).[68]

But he is also human, and less than a month before the

---

67. To the extent that Bland resembles any particular historical figure, it
might be C.D. Jackson, Vice-President of *Time*, publisher of *Fortune*, promoter of
Radio Free Europe, and close aide to President Eisenhower. The two Bland
organizations allude on the one hand to the Bland Corporation which financed
research into the feasibility of a 'Doomsday machine' in Stanley Kubrick's 1963
film *Dr. Strangelove*, and on the other to the RAND Corporation established in
1946 in Santa Monica, California, under the aegis of the Douglas Aircraft Corporation. See Blanche Wiesen Cook, 'First Comes the Lie: C.D. Jackson and Political
Warfare,' *Radical History Review* 31 (1984), pp. 42-70; Peter George, *Dr. Strangelove*
(London: Corgi, 1963), p. 98.

68. Roosevelt's position is discussed in Barton J. Bernstein, 'The New Deal:
The Conservative Achievements of Liberal Reform,' in Barton J. Bernstein (ed.),
*Towards A New Past* (New York: Vintage, 1969), pp. 263-88; William Appleman
Williams, *The Contours of American History* (1961; New York: New Viewpoints,
1973), pp. 439-69; van der Pijl, pp. 93-137 passim. See also H.G. Wells's contem-
porary estimate of FDR quoted in W.H.G. Armytage, *The Rise of the Technocrats*
(London: Routledge & Kegan Paul, 1965), p. 250.

German surrender Roosevelt dies. He is replaced by Harry Truman, a former haberdasher, bank clerk, West Point reject, and machine politician from Missouri – hardly Bland's ideal choice to maintain the 'synthesis.' But in the summer of 1945 it is not only Roosevelt who is dead. So too are Adolf Hitler, the figure who had facilitated the totalitarian transition in Europe, and Carl Duisburg, the General Manager of Bayer Chemicals who organized I.G. Farben after seeing John D. Rockefeller's Standard Oil in action in 1903, who used the term *'Führer* principle' long before Hitler was even heard of, and whose I.G. system constituted (in Wimpe's words) 'the model for the very structure of nations.' These deaths, along with those of Rathenau and Chicherin, raise the problem of succession and survival. The six solutions listed by Max Weber in his *Economy and Society* have some appeal, but all preserve this human frailty, this reliance on powerful leaders. The need remains for some final, permanent vehicle for the synthesis of human, natural, and technical resources, some stable operational motive beyond human vicissitudes (*GR* 349).[69]

This problem constitutes the rarely stated but crucial issue of post-war planning that runs throughout *Gravity's Rainbow.* For upon its successful resolution depends the salvation of 'Control' and the persistence of the Operation behind it. The solution, partially condensed in the phrase 'the routinization of charisma,' is sketched out early in the novel, not by Pointsman, who is already too involved in his own personal fantasies of control to care, but by Dr. Géza Rózsavölgyi, a Soviet anti-Pavlovian exiled in perpetuity to PISCES and The White Visitation, whose program for the unstructured, projective testing of patients claims to provide an irresistible and much deeper means of analysis than Pointsman's structured stimuli (*GR* 81-2). As the pressures on PISCES mount with the war nearing its end, Brigadier Pudding fading quickly, and Pointsman folding in on himself, Rózsavölgyi projects a way forward which maintains the Operation's control without the need for overt violence and without reliance on men at the center:

> The only issue now is survival – on through the awful interface of V–E Day, on into the bright new Postwar with senses and memories intact. PISCES must not be allowed to go down under the hammer with the rest of the bawling herd. There must arise,

---

69. See Yergin, p. 72; Borkin, pp. 5-6; Aron, p. 245.

and damned soon, able to draw them into a phalanx, a concentrated point of light, some leader or program powerful enough to last them across who knows how many years of Postwar. Dr. Rószavölgyi tends to favor a powerful program over a powerful leader. Maybe because this is 1945. It was widely believed in those days that behind the War – all the death, savagery, and destruction – lay the *Führer* principle. But if personalities could be replaced by abstractions of power, if techniques developed by the corporations could be brought to bear, might not nations live rationally? One of the dearest Postwar hopes: that there should be no room for a terrible disease like charisma ... that its rationalization should proceed while we had the time and resources. (*GR* 81)

Rószavölgyi seeks some project that organizes the total energies of society and consequently exemplifies it, as the construction of the wall does in Kafka's *The Great Wall of China* (1931); a program free of the senseless irrationalities of the Nazi past and capable of containing struggle just short of victory for an indefinite period. He will find it in the plot of *Gravity's Rainbow* itself, in the universal conspiracy to construct a rocket from stolen parts and to fire it, which constitutes Pynchon's analog of any process whose goal is synthesis – religious, philosophical, technical, historical – and inside which any act, however spontaneous in appearance, is in fact predetermined. For it is through this cruelly totalized process, centered on the V–2, that Pynchon's diverse materials are incorporated into a single universal plot, one which fulfils the manic dream of permanent location and absolute security, but which does so at the cost of paranoid anxiety inside an arc drawn, as if by gravity, towards suicidal waste: the dissolution at the end of the rainbow. And it is around the V–2, liberated from the rubble of the Oven State, that the new post-war synthesis – the Rocket State – coheres, mobilizing scientists and engineers, politicians and businessmen, soldiers, workers, spectators, and consumers, heroes and hero-watchers, systemizers and escapists in a drive for the absolute vantage-point. With the collapse of the primitive totalitarianism of Nazi Germany new movements develop which embody the internal history of the emerging Cold War: the dispersal of the supra-ideological technology of rocketry across the shifting ideologies of politics, and the appearance of men for whom rocketry has become ideology and idolatry.[70]

---

70. See Clarence Lasby, *Project Paperclip: German Scientists and the Cold War* (New York: Atheneum, 1971), pp. 194-5.

# 2

# No Man's Land

## The Osmosis of War

> The war, or rather *war*, was odd, he told himself a little inanely.
> But he knew what it meant. It was all covered with tedium and
> routine, regulations and procedure, and yet there was a naked and
> quivering heart to it which involved you deeply when you were
> thrust into it. All the deep dark urges of man, the sacrifices on the
> hilltop, and the churning lusts of the night and sleep, weren't all of
> them contained in the shattering screaming burst of a shell, the
> man-made thunder and light?
>
> — Norman Mailer, *The Naked and the Dead* (1948)

General Cummings' reverie on warfare on Norman Mailer's first
published novel, along with his journal notes on Spengler,
Nietzsche, and the social order required by organicist curves of
naturalized inevitability, are taken up twenty-five years later by
Thomas Pynchon in his investigation of a world increasingly
given over to a program of totalitarian engineering. But to
Mailer's fiction he adds two points. First, Pynchon proposes an
extension of the field of conflict following World War II. The
unanswered questions which confront Pointsman, Slothrop,
Roger Mexico, and others in *Gravity's Rainbow* – 'don't you know
there's a war on, moron? yes but ... where *is* the war?' (*GR* 54) –
already indicate the changing nature of warfare within the emerging incipient totalitarian state. The 'tedium and routine' of the
new warfare state, and the 'regulations and procedure' which
hold its units on permanent action stations, are no longer
presented by Pynchon as facets of some 'covering,' now detached
and left over from the receding transitional war recorded in
Cummings' reverie. Rather, they constitute mechanisms for that
conflict's absorption, transformation, and dispersal throughout

the Rocket State, which therefore extends in turn *as* a state of war. The 'naked quivering heart' of Cummings' transitional war is therefore dissipated within the social structure, a process indicated by the dispersal of the 00000 across the Rocket State. And the new privates of that state, the soldier–worker– consumers, necessarily find themselves 'thrust into it' continuously.[1]

This reorganization is recorded in a change of metaphor. Following 1945 the 'naked quivering heart' of World War II, which had been 'contained in the shattering screaming burst of a shell' in Mailer's novel, yields to the 'delta–t' of Pynchon's new war, which may manifest itself at any point. The war beyond the war develops, not as an imposition of front lines, rear areas, and safe havens, armed forces and non-combatants, tours of duty and furloughs, blitzes and all-clears, but as a condition of immanent tensions and perpetual competition in which violation and sacrifice are endemic, neutrality is meaningless, and engagements prove obligatory but inconclusive; a state where what were once discrete combat zones dissolve into one continuous no man's land, boundless in its domain but besieged at all times, ostensibly unrestricted and demobbed but potentially hostile at all points. Beneath the shadow of the delta–t, *Gravity's Rainbow* dramatizes an emergent society which will include not only men like Cummings but more importantly his civilian counterparts, descendants, and potential adversaries, all of whom will be on permanent active service.[2]

Second, and as part of his sense of a transformation in the nature of conflict, Pynchon includes the consequences of the 'silent rhapsodic swoop of the shell' beyond the mere sound of its explosion. Standing before the artillery piece in *The Naked and the Dead*, Cummings can still fantasize himself in a detached state of *Übersichlichkeit*: 'Just before he fired he could see it all ... The troops out in the jungle were disposed from the patterns in his mind ... all the violence, the dark coordination had sprung from his mind.' Discharging the weapon can still satisfy his sterile dream of violation as the wellspring of identity – 'in firing the gun

---

1. Norman Mailer, *The Naked and the Dead* (1948; St. Albans: Panther, 1964), pp. 477-82. See that portion of the 1955 Special Report of the US Defense Advisory Commission quoted in James Gilbert, *Another Chance: Postwar America, 1945-1968* (Philadelphia: Temple University Press, 1981), p. 96.

2. Mailer, *The Naked and the Dead*, p. 477.

he was a part of himself' – by allowing him to experience a 'deep boundless ambition' as well as a feeling of 'power ... beyond joy' which leaves him 'calm and sober.' But such license is already out of date. Following 1945, the maturation of the Rocket State means that at its core, military offensives may equal suicide, a condition finally acknowledged at the Geneva summit in July 1955. While Cummings' firing of the artillery shell marks an intermediate stage in that separation of responsibility and guilt between classical fascism and modern totalitarianism later recorded in *The Presidential Papers* (1963), the fantasies of power and security which breed in the resulting space remain precisely that: 'as foolish as shields of paper,' in the words of Oberst Enzian. In Mailer's first novel Cummings may survive but Hennessey does not. In *Gravity's Rainbow* Enzian extends the case within conditions he is himself helping to establish: 'the Rocket can penetrate, from the sky, at any given point. Nowhere is safe' (*GR* 728).[3]

In the terms of his journal General Cummings as gunner takes up a 'death viewpoint' in order to fantasize the sensations of a 'life viewpoint.' In that moment he feels secure. But this is the dream of a grown-up baby which transforms history into metaphors of inevitability, natural or mechanical, which requires a subordinate like Hearn to be 'nothing but a shell,' and which implies a society of would-be *Führers* engaged in limitless combat: a program of heroic vitalism and the antithesis of any sort of social coherence based on mutual aid or trust. Pynchon is haunted by the persistence of this mass fantasy of personal security whose grip reaches deeply into the sacrificial combat state: the desire, shared by so many of his own characters as by Cummings, to retreat into any totalizing structure which seems to offer protection – what Enzian terms 'shelter in time of disaster' – from the violent, the irrational, or the creative. In *Gravity's Rainbow* this totalitarian dream of inviolability within a single universal plot is dramatized as a recurring mechanism of human destructiveness – the 'virus of Death' – since it necessarily incorporates a nightmare of entropic waste, renders the human rigid and fearfully anxious

3. Mailer, *Naked and the Dead*, pp. 37, 476-8; Norman Mailer, *The Presidential Papers* (1963; St. Albans: Panther, 1976), pp. 191-2. See also Herbert Marcuse, *Negations*, trans. Jeremy J. Shapiro (1968; Harmondsworth: Penguin, 1972), pp. 263-4; Stephen E. Ambrose, *Rise to Globalism. American Foreign Policy, 1938-1976*, rev. edn (Harmondsworth: Penguin, 1976), pp. 240-44; Robert A. Divine, *Eisenhower and the Cold War* (New York: Oxford University Press, 1981), p. 122.

within its codings, and constitutes a neurotic mania whose realization may be suicidal. Moreover, Pynchon's epic implies that this may be an invariant condition in all human societies. The drive from uncertainty into structure, from mobility to stasis, from action to inertia — the conversion, in Pynchon's terms, of Zone into State — recurs throughout his fiction. This persistent drive, imaged through the construction and deliverance of Blicero's final rocket and child, survives beyond the dissolution of the Oven State to structure the nascent Rocket State that spreads beyond the launch of the 00000 (*GR* 723, 728).[4]

## Striking the Lyre

> It's a collection day, and the garbage trucks are all heading north toward the Ventura Freeway, a catharsis of dumpsters, all hues, shapes, and batterings. Returning to the Center, with all the gathered fragments of the Vessels.
>
> — Thomas Pynchon, *Gravity's Rainbow* (1973)

Before moving on to document the emergence of the new order in the post-war United States, we need first to summarize Pynchon's intimations of its overall structure and then to specify the fictional characters whose historical counterparts will execute what he, at least, suspects is its ultimate design. The Rocket State, its origins, range, domain, and bearing, as well as its justification of false mobility within naturalized security, are manifest both generally and particularly in the second half of *Gravity's Rainbow*, indicating the survival and reproduction of the 'virus' of totalitarianism. In a general sense it extends from the receding flood tides of World War II towards the closing firestorms of nuclear war; from Genesis to Revelation. It is thus produced and reproduced beneath the shadow of an 00000 whose rainbow trajectory satirizes God's covenant to Noah as the paradigm of hallucinatory safety harnesses: a false and useless promise of salvation to be betrayed by His own laws of nature. More precisely, it is inscribed between two converging parabolas, the first reaching from the UFA movie studio at Neubabelsburg in 1945 to the Orpheus Theater in Los Angeles in the present, the second reaching from the launch pad of the 00000 on Luneburg

---

4. Mailer, *Naked and the Dead*, pp. 69, 480-81.

Heath in 1945 to the same theater's roof at the point of its destruction. In describing the movement from the active production of a spectacle to the passive consumption of its products, the former accommodates that process by which Tyrone Slothrop's initial rehearsal of incipient totalitarian identities and his subsequent dismemberment lead to the partial production of Clive Mossmoon and other mass men and women from his scattered shells. In describing the absorption and dispersal of the 'virus of Death' throughout the incipient totalitarian state, the latter accommodates the related process by which Blicero's infection of Gottfried leads towards the consummation of Mossmoon's molded mass production.[5] The convergence of these two arcs records the production, dissolution, and reproduction of the postwar order's representative units through the interactions of labor and consumption, action and spectatorship, as intersecting processes defining the domain of the Rocket State.

Not only do these two arcs map out the Rocket State's general domain, however: they also specify its limits. For if their convergence accommodates that movement from the production of a spectacle to the consumption of its products which provisionally permits Mossmoon's and Cummings's other descendants in the incipient totalitarian state to believe at once in individual and personal power as well as in anonymous safety in numbers under unstated or benevolent authority, their intersection reveals such conceptions to be at best hallucinatory and at worst suicidal. On the one hand, the point at which the rocket touches the roof of the Orpheus Theater is the instant of its maximum velocity – it is 'falling nearly a mile per second' – and hence the apotheosis of the post-war integration of history and autobiography inaugurated by the flight of the 00000. In its public dimension it marks the climax of that naturalizing accommodation of 'violent political opposites' engineered within the rocket's parabola, since it completes the 'immachination' of its 'journey' into 'destiny' which takes hold beyond the *Brennschluss*; in its private dimension it marks the perfection of Clive Mossmoon's production techniques, since it is the site of the most rapid synthesis of his components and programming forged by the fastest of tools. But on the other hand this 'last unmeasurable gap' is also the terminus of the process: where the rainbows strike the surface and

---

5. On Mossmoon's production, see above, pp. 37–59.

disperse; where Mossmoon's means of production become the instruments of his death; where the absorption of 'the last delta–t' completes the integration process and with it the Rocket State's cycle. The absorption of private and public spaces into the singular plot of civil society may mark the consummation of the incipient totalitarian state, 'that city of the future where every soul is known and there is noplace to hide,' but it is also the mechanism and the occasion of its abolition (*GR* 318, 396, 566, 760).[6]

The volatile condition of the Rocket State is most clearly revealed in the final 'DESCENT' scene of *Gravity's Rainbow*, which describes 'the last delta–t' as a 'dark and silent frame' in which 'the pointed tip of the Rocket' descends on 'the roof of this old theater.' Inside, 'old fans who've always been at the movies (haven't we?)' and who have gathered to witness the arrival of a 'star' now clap and call for 'the show' to begin. But as they press forward for a glimpse of the star – 'Come-*on*! *Start*-the-*Show*! Come-*on*! *Start*-the-*Show*!' – so the rocket's re-entry leads these extending and stiffening members of the post-war state to be exploded: literally by the rocket, metaphorically by the contradictions embedded in their own productive cycle. In Goethe's *Faust*, the latter's reaching for the image of Helena in Mephistopheles' own 'spirit show' triggers a clouding of her shape and an explosion, whereupon Helena vanishes, the spirits dissolve, and the drama ends in 'darkness' and 'tumult'; in *Gravity's Rainbow*, Pynchon's 'old fans' clap and yell their way into a congruent void. They too accept the Mephistophelean key, the promise of access to 'the deepest, nethermost shrine' of existence, and they too fall victim to a catch in their contract: what Pynchon calls 'the terms of the creation.' Stated within the terms of the Rocket State, the theater-dwellers from Neubabelsburg to Los Angeles are first directed within and then sacrificed to a power structure they have themselves helped to establish. By accepting the contract for security within the sacrificial combat state, the lure of inviolability and personal power within a universal system of history, they fall into dependence and addiction as operators and victims of a propagandistic society. As the rocket reaches the roof of the theater, the motivation behind the totalitarian drive is revealed as a fraudulent and deteriorating condition. The star up

---

6. Cf. Johann Wolfgang von Goethe, *Faust, A Tragedy*, trans. Walter Arndt, ed. Cyrus Hamlin (New York: Norton, 1976), Part II, Act II, 8445-87. All subsequent references will be to this edition.

above is '*not a star*' but 'a bright angel of death,' the terminal and exterminating rocket (*GR* 413, 729, 760).[7]

Within the universal enclosure of the Orpheus Theater, Pynchon's *theatrum mundi*, its titular myth then assumes a pervasive presence, making withdrawal to some secure vantage-point beyond the underworld's reach, that dream of 'shelter in time of disaster' embraced by so many of the novel's characters, an anachronistic objective. Following 1945, neutrality proves untenable: in the words of William Burroughs, 'there are no innocent bystanders.'[8] So that when 'night manager' Richard M. Zhlubb – a thinly-veiled President Nixon – cruises in the engineered security of his 'Managerial Volkswagen' beyond the 'state of near anarchy' attending the midnight show, he is as much 'under the final arch' as his customers. The Hollywood Freeway is a part of the theater, and when the 'old fans' reach for their rising star, the face they think they know, the Rocket State's holding pattern starts to break down everywhere. Zhlubb ignores the warnings given by Heurtebise in Jean Cocteau's *Orphée*, even after his recent 'Bengt Ekerot/Maria Casarès Film Festival,' and 'looks up sharply into his mirror' at the sound of the siren. The universal 'rush hour' has come. Eurydice becomes your idiocy – 'words are only an eye-twitch away from the things they stand for' – and, as Orpheus puts down his harp, no more lyre becomes no more lies. The post-war order, overloaded like Mephistopheles by its own productions and demands, is 'reduced to telling truth at last.' Pynchon's final words absorb those of Mephistopheles at the point of explosion: 'It's your own work, this ghostly mask, you dunce!' And here the frame jams solid or the bulb burns out. The revelry is ended, the text is suspended, hung on what Pynchon calls the last 'CATCH' between illumination and extermination (*GR* 4, 100, 754-7, 760).[9]

---

7. Goethe, *Faust*, Part II, Act I, 6549-65, Act II, 7034. The shouts of Pynchon's theater-goers engage the Chamberlain's request to Mephistopheles at l. 6308: 'The master is impatient – start the show.' On Pynchon's use of the theatrical metaphor, see Charles Clerc, 'Film in *Gravity's Rainbow*,' in Charles Clerc (ed.), *Approaches to Gravity's Rainbow* (Columbus: Ohio State University Press, 1983), pp. 136-7; Marcus Smith and Khachig Tololyan, 'The New Jeremiad: *Gravity's Rainbow*,' in Richard Pearce (ed.), *Critical Essays on Thomas Pynchon* (Boston: G.K. Hall, 1981), p. 180. The prevalence of the figure, employed by writers from Plato to Balzac and Freud, is discussed in Richard Sennett, *The Fall of Public Man* (1976; Cambridge: Cambridge University Press, 1977), pp. 34-6, 40, 64, 108-10, 313.

8. William Burroughs, *Exterminator!* (London: Calder & Boyars, 1974), p. 94.

But a catch can also be a haul, and this one continues to fill and enlarge the Rocket State's nets within the grounds of the last delta–t. So that while it may be 'the end of the line' for the post-war order, its carriages go on piling up – what Pynchon calls 'the Vertical Solution' – and its walls go on breaking down, thus extending the no man's land and the domain of the 'virus of Death.' Equally, although neutrality may be an untenable stance after 1945, its narcotic vantage-points still proliferate in order to house Cummings' multiplying descendants: 'There is no way out' any longer, but the 'Evacuation still proceeds' (*GR* 3-4, 735). That the structure continues to thrive on the point of collapse is evident in the 'DESCENT' of *Gravity's Rainbow*. For where Faust once reached for the image of Helena in Mephistopheles' drama, now countless fans try to catch the 'last image [of] a human figure ... a face we all know' in Zhlubb's theater. The fans, the Rocket State's civilians, have polymerized beyond Mossmoon's initial synthesis to the extent that the Faust myth, and its dream of *Übersichlichkeit*, has become a pervasive condition of post-war society. Necessarily, and as its discrete manifestations scattered throughout the text serve to indicate, the face on the screen is equally polymorphous. It effectively constitutes an integral image of the Faustian masses' needs and alibis across the Rocket State: a palimpsest of explanatory and evasive systems; a comprehensive shelter for those adrift in the Zone 'where nothing is connected to anything, a condition not many of us can bear for long.' It functions as a universal joint or absorbing ideology holding incipient totalitarian society in harness (*GR* 434).[10]

But this joint, whilst it holds the expanding Rocket State together, does so by a set of increasingly precarious and explosive catches whose continued durability remains in doubt at the end of the novel. The central joint of the imperial order's apotheosis, the *Führer* principle, broke down in the ruins of Nazism: in the

---

9. Goethe, *Faust*, Part II, Act I, 6364, 6546; Jean Cocteau, *Orphée*, in *Three Screenplays*, trans. Carol Martin-Sperry (New York: Grossman, 1972), pp. 132, 168, 174-5. Maria Casarès played the princess, the guardian or 'Death' of Orpheus, in Cocteau's *Orphée*. Bengt Ekerot played death in Ingmar Bergman's *The Seventh Seal*. See Cocteau, p. 190. On the use of the terms 'rush hour' and 'holding pattern,' see *Gravity's Rainbow*, pp. 26, 501, 753.

10. The locations of the various faces in *Gravity's Rainbow* are: pp. 74, 92, 142, 222, 277, 358, 413, 501, 635, 670, 714, and 733. On ideology, see Hannah Arendt, *The Origins of Totalitarianism*, new edn (New York: Harcourt Brace Jovanovich, 1973), pp. 468-72.

last days of the Oven State Blicero retreated into 'his own space,' the archaic and private 'Kingdom of Lord Blicero,' and thus exploded the remaining nexus of imperial society. In the provisional days of the Rocket State that results the overly uniform system that had once exploded the imperial order is absorbed and dispersed as the interiorized pervasive condition of incipient totalitarian society: the massed fans occupy themselves by retreating into any number of privatized spaces or theaters, modernized reproductions of Blicero's 'personally frozen frame,' which, since they are generated and assimilated within the social structure, extend rather than dismember it. Hence the last joint appears more secure than its predecessor: more deeply embedded, with a grip that is harder to resist. But, as Faust discovered, such an appearance may be an elaborate – and fragile – fiction. For this joint also carries more weight and absorbs greater strain: the demands of a society of would-be *Führers*. Its failure or withdrawal may yet prove catastrophic.

## Bit Parts: Leading Man

> The three great figures that the machine has bred and trained up in the course of its development: the entrepreneur, the engineer, and the factory worker.
>
> — Oswald Spengler, *The Decline of the West* (1929)

> He has, so I have heard him say,
> Been born but half in some prodigious way,
> Of intellectual traits he has no dearth,
> But sorely lacks the solid clay of earth.
> So far the glass is all that keeps him weighted;
> But he would gladly soon be corporated.
>
> — Goethe, *Faust* (1832)

The structure provisionally supported by the joint, the Rocket State, is, however, not simply a static load but a dynamic product as well. Its means and relations of production, distribution, and exchange are outlined – partially, at least – as Pynchon describes the Zone. The fragmentation of Tyrone Slothrop and the projected development of his 'offshoots' into 'consistent personae,' which together indicate a rise in the number of workers, an extension in the division of labor, and therefore an increase in the accumulation of capital under the 'new dispensation,' constitute the first of many interlocking mechanisms fostering the Rocket State's post-war development. At the same time, Slothrop's Christ-like dismemberment denotes (via the devilish medium of the narcotic and aphrodisiac mandrake root) that sacrificial transformation of sexual energy into a bank of money – 'the alienated capacity of mankind,' 'the God among commodities' – which also played a central role in the extension of the incipient totalitarian order (*GR* 625, 742).[11] But this expansion of civil society into the deepest reaches of the private, the conversion of erotic Zone into 'universal whore' state, further includes the motivations translated by Enzian deep in his *Mittelwerk* prison camp 'Illumination' where his coupling 'with a slender white rocket' leads to 'dioramas on the theme "The Promise of Space Travel"' and a subterranean landscape of space suits and space helmets, 'oddly-colored television images,' and 'amusing little Space-Jockeys ... who will someday zoom about just outside the barrier glow of the *Raketen-Stadt*' in a 'strangely communal Waltz of the Future.' It includes, that is, a third mechanism of the Rocket State's growth: the creation of novelty and its systematic absorption as opportunity and necessity; the appearance of the charismatic and its routinization as commodity and spectacle, instanced here by previews of the nascent aerospace industry in its broadest sense. These processes are facilitated, in turn, not only by 'the bureaucracies of the other side' (headed by the late Walter Rathenau), but also by what Rathenau calls 'the technology of these matters,' specifically, the Schwarzkommandos' construction of the 00001, which constitutes a fourth term in the new order's means of production. For while Enzian believes that it 'may be possible to ride the interface ... all the way to the end between armies East and West' and thus to escape from the Allies' 'hardening occupation,' the Baltic is not the Red Sea and the rockets' historical destination remains Cuxhaven and 'Operation Backfire,' the Allies' first test firings of their newly acquired treasure. So that this new Moses delivers his followers from the Oven State only to lead them into its direct descendant, the

---

11. Cf. Karl Marx, 'Economic and Philosophical Manuscripts,' in *Early Writings*, intro. Lucio Colletti, trans. Rodney Livingstone and Gregor Benton (Harmondsworth: Penguin/New Left Review, 1975), pp. 285, 377; Karl Marx, *Grundrisse: Foundations of the Critique of Political Economy*, trans. M. Nicolaus (Harmondsworth: Penguin, 1973), p. 221; Oswald Spengler, *The Decline of the West, Vol. II. Perspectives on World History*, trans. Charles Francis Atkinson (London: George Allen & Unwin, 1929), pp. 505-7.

Rocket State. Both as seer and as leader, Enzian acts as an agent of the Operation by providing it with access to new and profitable colonization ventures: the 00001 becomes the first term in a new integration which will carry 'The Promise of Space Travel' towards its realization a generation beyond (*GR* 163-7, 296-7, 411, 727, 731).[12]

The multiple intersections of these and other elements create a field of potential relationships through which three figures – Franz Pökler, Clayton 'Old Bloody' Chiclitz, and Gerhardt von Göll – will increase their influence after the war. First, Franz Pökler's career takes off alongside the growing significance of the rocket inside the power structure as it mutates from its imperial to its totalitarian form. Closely duplicating the German years of Wernher von Braun, he moves from the *Technische Hochschule* in Munich to the *Verein für Raumschiffahrt* (German Rocket Society) in Berlin, and then to the German Army Ordnance Corps's proving ground at Kummersdorf, to Peenemünde, and finally to Nordhausen, where he constructs the Imipolex 'insulation device' for the 00000. He is, at least for a time, the rising technician, first cousin to William Burroughs' 'technical sergeant,' and considers himself 'a practical man' ('all things to all men, brand new military type, part salesman, part scientist') whose politics extend no further than the administration of his research funding. Although he dreams of moon flight he justifies it not as fantasy but as a means of transcending the politics and warfare of nations: his ideal community is given in Fritz Lang's 1926 film, *Metropolis*:

> a Corporate City-state where technology was the source of power, the engineer worked closely with the administrator, the masses labored unseen far underground, and ultimate power lay with a single leader at the top, fatherly and benevolent and just, who wore magnificent-looking suits and whose name Pökler couldn't remember.

And where class conflict is ultimately displaced by gilt-edged romance (*GR* 578). Although his decision to 'quit the game' in the dying days of the Oven State seems to distinguish Pökler from

---

12. Marx, 'Economic and Philosophical Manuscripts,' p. 377, calls money 'the universal whore, the universal pimp of men and peoples.' On the 'Backfire' project, see Frederick I. Ordway III and Mitchell R. Sharpe, *The Rocket Team* (London: Heinemann, 1979), pp. 294-309.

Burroughs' archetypal agent, who blows the world up because it is his job to do so, he first completes Gottfried's shroud, and his decision to 'quit' does nothing to stop his joining the next game, to be pulled in by new stakes and new players. Pökler remains 'just the type they want.' He or his colleagues will follow von Braun to the United States. (*GR* 154, 161-2; 219, 399-433, 578)[13]

Clayton Chiclitz constitutes a second essential agent of incipient totalitarian society who will rise to prominence following 1945: the standard parasitic business figure who intervenes in the changing power structure to advance his own position. After Pearl Harbor he translates military conflict and civilian anxiety into effortlessly marketable entertainment ('the enormously successful Juicy Jap ... realistically squishy plastic'), and then, as victory in Europe approaches, maneuvers himself into the war booty system in search of ways 'to cash in on redeployment' to the Pacific. By the summer of 1945 Chiclitz 'has his eyes on the future.' If the nature of that future remains unclear, his own thinking recognizes that capital thrives on conflict: 'Got to get capitalized, enough to see me through ... till we see which way it's gonna go. Myself, I think there's a great future in these V-weapons. They're gonna be really big.' Chiclitz' future is assured. A keen supporter of the Operation's leading figures – Hoover, Swope, Wilson – he is en route to succeeding them: in *The Crying of Lot 49* (1966) Chiclitz reappears at the centre of the military–industrial complex heading the vast Los Angeles-based Yoyodyne Inc., which has fed and grown on the Cold War system by producing high technology aerospace equipment and reproducing itself.[14] But twenty years before this moment, in *Gravity's Rainbow*, Chiclitz is already planning to extend his tentacles to the west coast. His dream of taking 'about 30 kids on [his] payroll ... back to America, out to Hollywood' – 'I think there's a

---

13. Von Braun attended the *Technische Hochschule* in Berlin. On his early career and the origins of the rocket's rise, see Ordway and Sharpe, pp. 12-29; Wernher von Braun and Frederick I. Ordway III, *The History of Rocketry and Space Travel* (London: Nelson, 1967), pp. 44-59; William Burroughs, *The Naked Lunch* (1959; London: Corgi, 1974), pp. 83-4; William Burroughs and Eric Mottram, *Snack* (London: Aloes Books, 1975), p. 8.

14. Thomas Pynchon, *The Crying of Lot 49* (1966; New York: Bantam, 1967), pp. 14, 59. On the 'T-Forces' which Chiclitz joins in order to get access to German military and technical capabilities, see Clarence Lasby, *Project Paperclip: German Scientists and the Cold War* (New York: Atheneum, 1971), pp. 18-26; James McGovern, *Crossbow and Overcast* (London: Hutchinson, 1965), pp. 99-105.

future for them in pictures' – fuses his cheap labor and diversific-
ation schemes in the form of a refugee welfare operation. The
scheme condenses the conditions of dependence presented as the
good life under capitalism, whilst Duane Marvy's comments on
life under Director De Mille – 'it's f'damn sure they ain't goam
be *singin'*. He'll use them little 'suckers for *galley slaves!*' – decodes
the image: the 'real big numbers, religious scenes, orgy scenes'
Chiclitz has in mind for the dependents will be staged at the
Rocket State's curtain (*GR* 558-9, 565).[15]

Such enterprises carry Chiclitz into the sphere of operations of
the third rising figure, Gerhardt von Göll, whose dream factory
career documents the inflection of propaganda from its imperial
form as a discrete intrusion on society to its totalitarian form as its
continuous and pervasive condition.[16] Von Göll's early work for
UFA at Neubabelsberg in the 1920s and 30s leads to his first
collaboration with Greta Erdmann on *Alpdrücken* and then to the
production of *Good Society*, a shift in titles indicating not only von
Göll's immediate need to placate the Nazi hierarchy after filming
the unreleased *Das Wütend Reich*, but also the coming official
dissolution of that twentieth-century nightmare into a structure
of communal benevolence. Then, with the collapse of the Oven
State, von Göll moves even deeper into the service of 'the new
dispensation.' For whilst *Alpdrücken* was shot 'at the height of his
symbolist period,' his later films are made predominantly 'on
location' using 'more natural light.' By the summer of 1945, he is
busy rushing round what is rapidly becoming one continuous
film set 'in a controlled ecstacy of megalomania' – 'his corporate
octopus wrapping every last negotiable item in the Zone' – trying
to carry out what he considers his last historic mission: 'to sow in
the Zone seeds of reality' (*GR* 388, 394-5, 611). This final and
continuing project under the aegis of the Rocket State is von
Göll's closest approximation, within technical and financial

---

15. On the links between Hollywood – including Cecil B. De Mille – and the
nascent aerospace industry, see Anthony Sampson, *The Arms Bazaar* (1977;
London: Hodder & Stoughton, 1978), pp. 92-6.

16. Marshall McLuhan, *Dew-Line Newsletter* 2, no. 3 (1969), poster 1: 'Propaganda does not consist in the conveying of messages by press or other media, but
consists in the action of the total culture … upon its participants. The idea that
propaganda consists of packaged concepts peddled to unsuspecting citizens is no
longer tenable.' This point is most thoroughly discussed in Jacques Ellul, *Propaganda: the Formation of Men's Attitudes*, trans. Konrad Keller and Jean Lerner
(1968; New York: Vintage, 1973), especially pp. 118-21.

limits, to his dream of a propagandized and narcotic society beyond perception, discussion, or evasion; a totalitarian control structure outside history – at once film set and movie theater, make-up department and cutting room, box office and casting couch – from which Joyce's Stephen Dedalus would not think to awake:

> There is a movie going on, under the rug. On the floor, 24 hours a day, pull back the rug sure enough there's that damn movie! A really offensive and tasteless film by Gerhardt von Göll, daily rushes in fact from a project which will never be completed. Springer just plans to keep it going indefinitely there, under the rug. The title is *New Dope*, and that's what it's about, a brand new kind of dope that nobody's ever heard of. One of the most annoying characteristics of the shit is that the minute you take it you are rendered incapable of ever telling anybody else what it's like, or worse, where to get any. Dealers are as in the dark as anybody. All you can hope is that you'll come across somebody in the act of taking (shooting? smoking? swallowing?) some. It is the dope that finds *you*, apparently. (*GR* 745)[17]

But von Göll's corporate octopus can no more monopolize the conversion of Zone into State than can Chiclitz' nascent Yoyodyne. Nor will Pökler's dream of technological determinism be realized within the rocket's domain. For after 1945 'ultimate power' no longer lies 'with a single leader at the top' – justice, benevolence, and an impressive wardrobe notwithstanding – nor with a single instrument or program. Rather, as a result of Blicero's cession of power in the Oven State's clearing, authority is transferred to the synthetic complex of 00000, Imipolex, and Gottfried, and then delivered across the post-war world in the form of a 'virus of Death' which spreads the plague of totalitarianism into Mossmoon and his masses, into the flexible and fetishistic commodity bonds created by the 'material of the future,' and into the descendants of the V–2 itself. It becomes dispersed, that is, throughout the entire Rocket State, whose continuing monolithic operation thereby assumes the mantle of authority. As a result, men like Pökler, von Göll, and Chiclitz, as well as their colleagues and competitors, are maintained not as masters but as *agents* of that structure, fighting for control of its networks

---

17. On the film industry and realism, see Brian Lee, *Hollywood* (Brighton: British Association of American Studies, 1986), p. 10.

yet dependent on them, reaching for the top only to be passed over, sooner or later, like Blicero and Pointsman before them (*GR* 488, 578).[18]

Nevertheless, the dispersal of the 'virus of Death' does not so much eliminate its sources and vectors as obscure them. The paths of the Rocket State's 'Typhoid Marys' can still be traced, even when buried in 'a ground of terror, contradiction, [and] absurdity.' Thus, one particular type of carrier, previously absent from Pynchon's European theater and to be added to the three figures previously discussed, is divined in the Oven State's valediction. For the casting of Blicero's tarot near the end of *Gravity's Rainbow* ('better than Slothrop's,' not surprisingly) indicates a successor to the throne of the former's kingdom in the shape of a 'fair intellectual king' whose tracks may be found 'among the successful academics, the Presidential advisers, [and] the token intellectuals who sit on boards of directors.' This modern king will not re-enact Blicero's old part to the letter, nor would such an endeavor be possible: as the structure of control changes so the personal monopolization of power once enjoyed by a *Führer* no longer survives. For the same reason he may not even take the part as a single figure. But he will not reject it completely either. For Blicero is not entirely passed over: he remains as 'the father you will never quite manage to kill,' the charismatic who, like the rocket itself, will never quite be routinized. Instead, this new king will give a modernized performance. Just as the incipient totalitarian order absorbs the imperial, so he will absorb the surviving father – not to depose him but to 'impersonate' him. Consequently, he will reappear on the post-war stage not as a figure of 'ultimate power' but, in innumerable guises, as the Rocket State's 'leading man' (*GR* 582, 746-9).

The performance the new king stars in will be directed by Gerhardt von Göll, produced and promoted by Clayton Chiclitz, and engineered by Franz Pökler. Its production will require the labor, skilled and unskilled, broken in by the dispersal of Tyrone Slothrop; the money that conditioning releases; the hardware and software, props and backdrops, designed to build on the rocket's first appearance; the motivation of sensational reviews; and the involvement at all points, active and passive, of millions of mobilized Mossmoons. It will draw on everybody in the theater. Each will give his or her services and each will take a cut,

---

18. On this dispersal, see above, pp. 54-9.

one way or another, in a new marriage of convenience: a rocket romance replacing the wartime lovers' shotgun wedding, now dissolved from alliance to mutual slanging. Pynchon's textual preview breaks off at this point, as 'the show' is about to begin, but it contains sufficient evidence to anticipate the plot. For as the parabolas leading from the UFA studios at Neubabelsburg and the Luneburg Heath clearing converge at the Orpheus Theater, creating the complex spatial and temporal intersection at 'the last delta–t' where the Faustian Rocket State is produced and consumed, so Pynchon projects a third parabola resulting from their occlusion. This arc is activated and directed within the boundaries of the delta–t and translates the potential energies of the closing parabolas' interactions into one specific kinetic which describes both the apotheosis and the limit of the Rocket State.[19] It gives an indication of what is produced inside the last delta–t by those means of production drawn together in its domain; of what sort of operation maintains and extends the incipient totalitarian order's fragile integration. Together, the 'oddly-colored television images' found deep in the *Mittelwerk*, the V–2 engineer's daughter's dreams of 'fly[ing] someday to the Moon,' Clayton Chiclitz's sense of 'a great future in these V–weapons,' Enzian's salvation of the first post-war rocket, and Pynchon's continued use of Goethe's *Faust* to associate the rising Rocket State with an incipient Apollo, all certainly suggest that the content of this third parabola will be the American-manned space program. But its course can only fully be charted beyond the pages of *Gravity's Rainbow* through the political, economic, technical, and cultural minutiae of post-war American society. The arc does not, after all, appear miraculously: it is itself produced, textually and historically.[20]

Textually, the rising curve is produced as the two converging parabolas occlude at the Orpheus Theater. It is therefore generated by the growing interaction of dramatic illusion and substantive action beneath the delta–t (proposed in Pynchon's use of *Faust*) and indicates the expansion in real life of what Nathaniel Hawthorne once gave in his introduction to *The Scarlet Letter* as the romantic locus of imaginative composition: 'a neutral territory between the real world and fairy-land, where the actual and

---

19. Cf. Mailer, *Naked and the Dead*, p. 275.
20. See especially the retreat of Faust and Helena and their production of Euphorion, the nascent Apollo, in Goethe, *Faust*, Part II, Act III, 9586-10038.

the imaginary may meet.' The immediate historical context of this process is the 'multivocal and polysemous' environment of modern capitalist and state capitalist societies: the world described in Jacques Ellul's *Propaganda*, immersion in whose 'universe of images' without sufficient information or discriminatory analysis leads to a blurring of definitions of life and fantasy – as Slothrop discovered at Neubabelsburg. Hawthorne's 'neutral territory' may thus be restated, following 1945, in a way Hawthorne never imagined, as a no man's land where both 'actual' and 'imaginary' are condensed into a single area of hallucination and fabulation: in Pynchon's words, a 'nonstop revue' with 'Outside and Inside interpiercing one another too fast, too finely labyrinthine, for either category to have much hegemony any more.' It is to the mechanics of this no man's land's production, first in *Gravity's Rainbow* and then in the United States itself, that we now turn (*GR* 681).[21]

## The Vertical Solution

'Springer, this ain't the fuckin' *movies* now, come on.'
'Not yet. Maybe not quite yet. You'd better enjoy it while you can. Someday, when the film is fast enough, the equipment pocket-size and burdenless and selling at people's prices, the lights and booms no longer necessary, *then* ... then ...'

— Thomas Pynchon, *Gravity's Rainbow* (1973)

... in the marginal area, the gap, were the peculiar tensions that birthed the dream.

— Norman Mailer, *The Naked and the Dead* (1948)

The fabulous statistics continued to pour out of the telescreen. As compared with last year, there was more food, more clothes, more houses, more furniture, more cooking-pots, more fuel, more ships,

---

21. Hawthorne, as quoted in Eric Mottram, 'The Location of Dangerous Shoals: American Fictions on the Science of Power,' *Over Here* 3, no. 2 (1983), p. 4; Victor Turner, 'Themes in the Symbolism of Ndembu Hunting Ritual,' in John Middleton (ed.), *Myth and Cosmos: Readings in Mythology and Symbolism* (1967; Austin and London: University of Texas Press, 1976), p. 254; Ellul, pp. 87, 144-5; Jacques Ellul, *The Political Illusion*, trans. Konrad Kellen (New York: Alfred A. Knopf, 1967), p. 112.

more helicopters, more books, more babies – more of everything
except disease, crime, and insanity. Year by year and minute by
minute, everybody and everything was whizzing rapidly upwards.

— George Orwell, *Nineteen Eighty-four* (1949)

Gerhardt von Göll's *New Dope* includes amongst its 'daily rushes'
the America presented in Charles Walters' *High Society* (1956), a
central dream of white aspirations in the 1950s in which the hard-
bitten working life and gossip-column cynicism of Frank Sinatra
and Celeste Holm dissolve before the true love of Bing Crosby
and Grace Kelly ('Miss Frigidaire') inside a universe of commodi-
fied and automatic ease, of 'swell parties,' and of sensation
permitted within a fixed social structure effortlessly accommodating both Crosby's 'gentle folk of Newport' and Louis Armstrong's
entertaining trumpeter.[22] As another profitable instance of Hollywood's post-HUAC policy of avoiding controversy – Satchmo
serenaded the wedding of 'gentle folk' while 102 Congressmen
signed the Southern Manifesto and the NAACP was banned in
Alabama – Walters's film equally constitutes a suitable sequel to
von Göll's *Good Society* (*GR* 394). But while *High Society* rehearses a
euphoric and plentiful culture, its title also indicates one that is
at once narcotic, dependent, precarious, and remote. And it is
this tensed enclosure – an American Dream surrounding or
warding off an immanent nightmare – which Pynchon condenses
as the 'intimate cubic environment' of the rising elevator near the
end of *Gravity's Rainbow*:

> By now the city is grown so tall that elevators are long-haul affairs,
> with lounges inside: padded seats and benches, snack bars,
> newsstands where you can browse through a whole issue of *Life*
> between stops. For those faint hearts who first thing on entering
> seek out the Certificate of Inspection on the elevator wall, there are
> young women ... who've been well-tutored in all kinds of elevator
> lore, and whose job it is to set you at ease. (*GR* 735)

Where *High Society* concludes with an image of terminal satisfaction not always repeated in American life – Grace Kelly may have
become an elite icon to the extent of being televised marrying

---

22. The corresponding conditions for wealthy society in mid-1950s America
are given in Gabriel Kolko, *Wealth and Power in America* (New York: Praeger, 1962),
pp. 133-4.

European money after filming, but Crosby relied on alcohol and proved a less than perfect parent – Pynchon's elevator satirizes the post-war media marketing world of upward mobility as a pervasive but treacherous routine of consumable security. The Rocket State becomes an environment located, like Henry Miller's *Air-Conditioned Nightmare* (1945), somewhere between insane asylum and shopping center.[23]

But the elevator, like the turning face, takes on many forms in *Gravity's Rainbow* and moves at the intersection of more than these specific terms and instances – dream and nightmare, restaurant and padded cell – of incipient totalitarian society. Indeed, in the course of the text its mobility and heterogeneity

---

23. Charles Walters, dir., *High Society*, Metro-Goldwyn-Mayer, 1956. On Hollywood after the HUAC hearings, see Richard Maltby, *Harmless Entertainment: Hollywood and the Ideology of Consensus* (Metuchen, N.J.: Scarecrow Press, 1983), pp. 118-30; Larry Ceplair and Steven Englund, *The Inquisition in Hollywood: Politics in the Film Community, 1930-1960* (Garden City, N.Y.: Anchor Doubleday, 1980), pp. 248-98, 325-45; Les K. Adler, 'The Politics of Culture: Hollywood and the Cold War,' in Robert Griffith and Athan Theoharis (eds), *The Specter: Original Essays on the Cold War and the Origins of McCarthyism* (New York: New Viewpoints, 1974), pp. 240-58.

24. Thus at 'the end of the line' the elevator transports evacuees into 'some vast, very old and dark hotel, an iron extension of the track and switchery by which they have come here.' But traversing frictionless passages it connects this current accommodation with a 'very extensive museum,' a memory bank of 'many levels and new wings that generate like living tissue,' and at the same time (in the form of a 'mobile building') gives access to 'spectacles' of anticipation staged in a 'dingy little amphitheater surrounded by 'hundreds of thousands of . . . *spectators*, watching . . . to see if a new episode's come on yet.' Spatially, the elevator incorporates both the 'great arena' and the 'intimate environment.' Dynamically, it moves, sometimes 'without friction' – 'zipping along the corridor-streets of the *Raketen-Stadt*' – and sometimes at a 'Slow Crawl' – 'depending on a secret process among the granters of permission.' Motivationally, it links a system of accession and plentiful ease, where 'numberless shelves, each one revealing treats gooeyer and sweeter than the last' invite you to '*go inside*,' with one of ominous restriction in which 'certain paths aren't available to you,' levels are 'somehow forbidden,' and halls are 'to be entered at one's peril.' Prospectively, it is engineered at the edge of present security and potential disaster: climbing 'window to window, too full of grace ever to fall' but at the same time 'propelling you with no warning toward your ceiling'; offering 'padded seats and benches' but also hanging like 'a moving wood scaffold' in which furnishings no longer matter; presenting 'a whole issue of *Life* between stops' yet already at 'the end of the line' where 'bricks and mortar showering down' may bring 'sudden paralysis as death comes to wrap and stun' (*GR* 4, 49, 537, 674, 679, 735).

If the face on the screen of the Orpheus Theater becomes the latest expression of Arendt's totalitarian ideology, the elevator network becomes its locus of terror, a novel form of concentration camp. See Arendt, *Origins*, pp. 465-8.

demonstrate that the no man's land is as universal as the
Orpheus Theater's Imipolex-based celluloid joint.[24] Within its
walls Pynchon's elevator condenses the Rocket State's structure
of time and space, power and mobility, mediation, propaganda,
satiation, and routine into the everyday life of post-war Fortress
America: powerful but vulnerable, cosy yet thrilling. But if the
elevator is a tensed network of contradictory and fragmenting
forces it is also the site of their apparent resolution: 'where ideas
of the opposite have come together, and lost their oppositeness.'
The mechanism of this conjunction is that of Mossmoon and the
Rocket State itself: mobility. The 'revue,' in Pynchon's words, is
'nonstop.' Supported within the Rocket State's scaffolding by the
straining pulleys and coiling spokes of bourgeois individualism,
bourgeois love, alienation, and endlessly reflective spectatorship,
the elevator moves continuously within the limits of the delta–t,
at once differentiating the 'crystal palace' into 'separate rows,
aisles, exits, [and] homecomings,' and integrating its 'thousands
of ... hushed [and] invisible rooms' into the single auditorium of
the Orpheus Theater. Constantly negotiating and reformulating
the relationships between the private and public dimensions of
the Rocket State, it constitutes 'an interface between one order of
things and another': between the tenuously integrated shells of
Clive Mossmoon, between Pain City and Happyville, and
between the terms defining those relationships: conscious and
unconscious, isolation and community, reality and imagination,
peace and war. Constantly relocating the boundaries between
Zone and State, it defines the reaches and limits of incipient totalitarian society (*GR* 3-4, 49-50, 198, 206, 302, 450, 663).

Yet this process of 'building or destruction under way in
various parts of the City,' Spengler's *perpetuum mobile*, does not so
much resolve the contradictions of the Rocket State as engineer
their containment. For just as the polymorphic face on the screen
of the Orpheus Theater constitutes a universal joint holding
incipient totalitarian society in harness, so the continuous
movement of the elevator organizes the theater's thousands of
'invisible rooms' into an endless tesselation of 'peep-show
machines' articulated through and around it. And whilst this
places the greater strain of an expanding Rocket State on the
joint, it also distributes that load across the entire structure of the
theater. Thus, where the central nexus of the imperial order's
apotheosis, the *Führer* principle, stiffened and broke down in
1945, now the maturation of the post-war order overhauls and
reactivates that juncture, not simply as its singular pivot but as its

pervasive condition. The contradictions of advanced capitalist society are therefore not resolved but absorbed by thousands of intersecting catches; they are contained by dispersal. In effect, the continuous movement of the elevator and the polymorphism of the face on the screen interact as two moments of a single process which maintains the tensed and tenuous boundaries of the Rocket State – the delta–t between the falling rocket and the Orpheus Theater – by perpetually adjusting its interior structure (*GR* 450, 725).[25]

In fostering this transformation, the moving elevator becomes the vehicle of the 'old fans'' retreat from the power structure analyzed in Wilhelm Reich's *The Mass Psychology of Fascism* (1933) into the privatized spaces implied by the title of David Riesman's *The Lonely Crowd* (1950).[26] It carries its passengers – General Cummings' multiplying descendants – towards a reorganized post-war security system of personalized vantage-points: sanctuaries where all viewers reside inside a private 'stall, each with [its] own key and locker, pin-ups and library shelves decorating the partitions ... even one-way mirrors so you could sit at your ease.' But this 'Evacuation' is 'no way out,' nor barely 'shelter in time of disaster.' For all the polite 'elevator lore' passed around the joint by Mindy Bloth and her zoot-suited tutors, these new enclosures remain as precarious – for oneself and 'all the others pressed in around' – as any society based on Darwinian competition and unlimited individualism. To the question 'no one dares ask, not out loud' – 'Is this the way out?' – Pynchon gives an answer which the rest of *Gravity's Rainbow* exhaustively documents: 'No, this is not a disentanglement from, but a progressive *knotting into* (*GR* 3-4, 450, 728, 735).[27]

The movement is progressive in that it bolsters and defends outmoded and destructive forms of social organization in the

---

25. Spengler, p. 502. Pynchon's 'City' is, in a number of ways, closely related to that total environment dramatized in the short story 'The Concentration City' in J.G. Ballard, *The Disaster Area* (1967; London: Panther, 1969), pp. 31-54. For a concrete expression see Murray Bookchin's analysis of American urban growth in *The Limits of the City* (New York: Harper & Row, 1974), especially, pp. 80-81.

26. See Otto Kirchheimer, 'Private Man and Society,' in Frederick S. Burin and Kurt L. Shell (eds) *Politics, Law and Social Change: Selected Essays* (New York: Columbia University Press, 1969), pp. 463-77.

27. Again, for concrete instances of such installations and their meaning, see Bookchin, pp. 81-6, and Lewis Mumford, *The City in History* (1961; Harmondsworth: Penguin, 1966), pp. 582-4. See also Ellul, *Propaganda*, pp. 8-9, 147-8.

course of propounding their replacement. It is a 'knotting into' in that its extension of the Rocket State involves and requires the concomitant enticement, conditioning, and exploitation of its executives. In both cases the historical context is the vaguely ameliorative politics of liberal capitalist reform in America and elsewhere whose human instrument is that synthesized prototypical unit of post-war society, Clive Mossmoon.[28] For under the guidance of Sir Marcus Scammony, Mossmoon becomes not only the representative agent of that post-imperial, post-laissez-faire alliance of business and state power through which the reforming Operation emerges victorious in 1945 to administer the system, but also its representative subject. Since he works for and advances within a reorganized control structure which is less a means of resolving the contradictions of post-war capitalism than of containing their instability, he at the same time incorporates the consequences of that dispersive and evasive action. At once executive and bureaucrat, expert and patient, governor and cog, Mossmoon imposes, contains, and suffers the fragmenting conditions of incipient totalitarian society; he embodies its unresolved intersections of stasis and anarchy.[29]

As the template for the molding of the Rocket State's masses, Mossmoon's post-war polymerization spreads this unstable nexus – generating both the manifold and protean isolated units of the lonely crowd and their articulating structure of load-bearing catches – and thus synthesizes the Rocket State as a precarious edifice manifesting immanent contradictions. At the same time, however, as an agent of the reforming Operation his reproduction also redevelops the post-war structure as an expansive and absorbing version of Marcuse's one-dimensional society: one, that is, where politics is converted into administration, where critical autonomy is dissolved into behavior fit for bourgeois psychoanalysis, and where, therefore, the forms in which these contradictions may be expressed are restricted. Consequently, Mossmoon's mass production constitutes and advances the customary liberal 'solution' to – in other words its

---

28. The American roots of such reform – and thus of reformers like Lyle Bland – are analyzed in James Weinstein, *The Corporate Ideal in the Liberal State: 1900-1918* (Boston: Beacon Press, 1968).

29. See Bookchin, p. 83, and Arendt, *Origins*, p. 468. Cf. Spengler, p. 505: 'The centre of this artificial and complicated realm of the Machine is the organizer and manager.'

expansive reinforcement of – the unresolved problems of capitalism: it extends their field by restricting their methods and
prospects of solution to those sanctioned and provided by the
Operation itself – that is to say, to those which capitalism may
use. Like the power structure dramatized in Joseph Heller's
*Catch–22* (1961), it proposes answers to everything while in fact
delivering solutions to nothing.[30]

But liberalism is no more than the adopted mechanism – and
Mossmoon simply an agent and victim – of the Operation, not
its objective. The latter only manifests itself incrementally as the
new 'State [which] begins to form in the Stateless German night'
extends following 1945. This process is, however, neither singular
nor enforced, but polymorphic and indulgent. Whereas in *Gravity's Rainbow* the ends of the imperial order are condensed in the
collapsing figure of Dominus Blicero, the apotheosis of the *Führer*,
those of incipient totalitarianism are dispersed for democratic
recovery across the entire transitional Zone. When Tyrone Slothrop embarks on his travels in the spring of 1945, what he
uncovers is not a single figure but a limited set of terms and
contacts within an indistinct, undisclosed, and heterogeneous
program; his discoveries are the result, or so he initially believes,
not of external manipulation but of individual labor, reflection,
and composition. And it is from the evidence Tyrone unearths as
well as from other key precipitates of the Oven State's dissolution
– Oberst Enzian's 00001, 'The Promise of Space Travel' recorded
by Fritz Lang's *Die Frau im Mond* and in other image banks – that
Pynchon adumbrates his exemplary expression of the
Operation's 'Next Higher Assembly,' the 'soul' of incipient totalitarian society: 'the Rocket' itself (*GR* 252, 296-7, 566).

## Things to Come

We are in a period now I think of the formation of a mood.

— Dean Acheson (December 1947)

Preparedness isn't all military.

— Robert Heinlein, *Destination Moon* (1950)

If the rocket escapes the Oven State as the technical product of
military necessity and as a celluloid spectacle for civilian gratific-

ation, it re-enters the no man's land of post-war society at their deepening intersection. During 1946 and 1947, Wernher von Braun began work on a fictional depiction of space travel, *Das Marsprojekt*, initiated test launchings of recovered V−2 rockets at the White Sands Proving Ground, New Mexico, and inserted mathematical proofs of the feasibility of space travel derived from the latter as an appendix in the former. During the same period the US Air Force's recently established Project RAND completed a report which found earth satellites such as the one recently proposed by the Office of Naval Research to be technically viable, and Robert Heinlein published the first of his best-selling space novels, *Rocketship Galileo*. Each event constituted a vital term in the Rocket State's gradual integration; together they indicated the extension of that region of invention and imagination 'where science is fiction and science fiction at any moment may turn into science.'[31]

But the 'transmarginal leap' between the actual and the imaginary can include a 'surrender' to hallucination, manipulation, and control, depending on the historical context: the terms of the 'nonstop revue.' In 1947 the no man's land took the form of a society stiffening under the imposition of the pervasive Cold War structure: what Allen Ginsberg later termed the Syndrome of Shutdown. That year the President announced the Truman Doctrine and State Department officials explicitly formulated the domino theory; George Kennan published his thoughts on containment, and George Marshall announced his European aid program; the House Un-American Activities Committee extended its anti-communist offensive into the film industry, and the Motion Picture Association of America began suspending employees and introduced a blacklist; Congress passed the Taft−Hartley Act and the Justice Department issued its first list of

---

30. Cf. Hannah Arendt, *The Human Condition* (Chicago: Chicago University Press, 1958), pp. 22-78; H.T. Wilson, *The American Ideology: Science, Technology and Organization as Modes of Rationality in Advanced Industrial Societies* (London: Routledge & Kegan Paul, 1977), pp. 171-99; Kostas Axelos, *Alienation, Praxis, and Technē in the Thought of Karl Marx*, trans. Ronald Bruzina (Austin and London: University of Texas Press, 1976), pp. 89-109; Maltby, pp. 244-51; Herbert Marcuse, *One Dimensional Man* (1964; London: Abacus, 1972), pp. 9-103.

31. Ordway and Sharpe, p. 361; Enid Curtis Bok Schoettle, 'The Establishment of NASA,' in Sanford A. Lakoff (ed.), *Knowledge and Power: Essays on Science and Government* (New York: Macmillan/The Free Press, 1966), p. 166; Erik Bergaust and William Beller, *Satellite!* (London: Scientific Book Club, 1957), pp. 34-5; Mottram, 'Location of Dangerous Shoals,' pp. 3, 15.

subversive organizations for the use of private industry; Truman ordered and extended the Federal Employee Loyalty Program to cover eight million people and signed into law the National Security Act creating the National Security Council, the Central Intelligence Agency, and the National Military Establishment – the command structure for the Cold War.[32]

The rocket therefore re-enters as a multistaged vehicle – at once product, project, and proposition – within a tesselated theater whose customers are themselves simultaneously mobilized and restrained short of victory or defeat in a tensed and tenuous space between war and peace. It is incorporated as a mobile point inside a field of propaganda whose terms polarize the viewers' vantage-points into an alternating structure of anxiety and relief – between security and invasion, purging and infection, enclosure and collapse, reinforcement and retreat, immunity and exposure – characteristic of post-war organizational psychology. The process of absorption is thus mutual: as the rocket re-enters so the viewers may be taken in. It is at the same time expansive: where Slothrop's Zonal quest once drew him via engineered hallucination towards the post-war State, his scattered shells now follow. 1947 also saw the first 'flying saucer' enter the American imagination. Between that June and December 1965, the US Air Force documented 10,147 reported sightings of Unidentified Flying Objects. By the late 1960s an annual convention at Giant Rock, California, was drawing over 10,000 enthusiasts.[33]

Within this field public enthusiasm may be liberally converted into private gain. For if constant preparedness maintains anxiety it also feeds a desire for relief: sooner or later – 'Come-*on*! *Start-*

32. See John Tytell, *Naked Angels: The Lives and Literature of the Beat Generation* (New York: McGraw-Hill, 1976), p. 5; John Lewis Gaddis, *The United States and the Origins of the Cold War, 1941-1947* (New York: Columbia University Press, 1972), pp. 316-52; Daniel Yergin, *Shattered Peace: The Origins of the Cold War and the National Security State* (1977; Harmondsworth: Penguin, 1980), pp. 275-336; George Kennan, 'The Sources of Soviet Conduct,' *Foreign Affairs* 25 (1947), pp. 566-82; Ceplair and Englund, pp. 254-98, 325-60; Richard M. Freeland, *The Truman Doctrine and the Origins of McCarthyism* (New York: Alfred A. Knopf, 1972); Thomas H. Etzold, 'American Organization for National Security, 1945-1950,' in Thomas H. Etzold and John Lewis Gaddis (eds), *Containment: Documents on American Politics and Strategy, 1945-1950* (New York: Columbia University Press, 1978), pp. 8-23.

33. See Robert Plank, *The Emotional Significance of Imaginary Beings* (Springfield, Ill.: Charles Thomas, 1968), pp. 123-40; Leslie Peltier, 'UFO,' in Arthur C. Clarke (ed.), *The Coming of the Space Age* (1967; London: Panther, 1970), pp. 292-5.

the-*Show*!' – the viewers want action. And when the fortified need to be exercised, the Operation will follow and then supply the lead. In 1951, his production of *When Worlds Collide* almost complete, George Pal began work on a film of H.G. Wells's *The War of the Worlds* (1898). The reception of Orson Welles's 1938 radio broadcast in his mind, Pal transferred the setting of the novel to present-day California, partly to obviate expensive sets and costumes but primarily to capitalize on the continuing public response to flying saucer stories following the first sighting over Washington State in 1947. Along with director Byron Haskin he felt that a contemporary and local setting would ensure the desired effect – what Haskin termed 'excitement.' Since Pal had already proved a good investment – *When Worlds Collide* made substantial profits – Paramount agreed and, drawing on the financial resources of the Bank of America, the Chemical Bank, and others, backed his production to the tune of $1 million (*GR* 760).[34]

As the original presentation for Pal's previous film revealed, a mass need for apocalyptic sensation constituted a profitable opportunity for a film industry fighting a losing battle against legal rulings, the spread of television, and resulting changes in investment patterns.[35] Noting the existence of 'pulp magazines with millions total circulation devoted exclusively to space travel, world disaster, etc.,' the backers of *When Worlds Collide* itemized the potentials of what one critic later termed 'the imagination of disaster':

> End of world is biggest conceivable spectacle ... It would show ... stupendous geological spectacles: earthquakes, volcanoes, hundred-feet tides, collapse of cities, inundation of cities, smashing of kitchens [sic] ... the people of the earth – some in terror – some in religious resignation – some struggling by science to escape – but most in a terrific orgy of rape, lust, robbery, murder, etc.

Such indulgencies, the scenario hedged its bets, would provide

---

34. Rudolph Mate, dir., George Pal, prod., *When Worlds Collide*, Paramount, 1951; Byron Haskin, dir., George Pal, prod., *The War of the Worlds*, Paramount, 1953. See John Brosnan, *Future Tense: The Cinema of Science Fiction* (London: MacDonald & Jane's 1978), p. 90; Bill Warren, *Keep Watching the Skies! American Science Fiction Movies of the Fifties. Vol. 1: 1950-1957* (Jefferson, N. Cal. and London: McFarland, 1982), p. 153; Janet Wasko, *Movies and Money: Financing the American Film Industry* (New Jersey: Ablex Publishing Corp., 1982), p. 185.

35. See Maltby, pp. 63-70.

both 'appalling background and suspense for love story' and a 'magnificent picture opportunity in take off of huge rocket.' Moreover, while the earth faced the prospect of being 'a. Dropped into sun; b. Hurled into space; [or] c. Dashed against another planet,' Paramount could look forward to the prospects of a low overhead, high profit sequel.[36]

The advocates of *When Worlds Collide* were reaching into the 'invisible rooms' of the post-war theater ('feeling for a point of intersection,' as William Burroughs puts it in *Nova Express*); loading, reloading, and overloading the 'peep-show machines'; unloading the takings and relieving the taken in; investing and withdrawing; capitalizing and expanding – acting as advance agents of the Rocket State's invasion of the diminishing Zone. Necessarily they depended for the film's financial success on prior accumulation, most recently the Operation's own five-year plan of investment in the Cold War image bank: the newest wing of their 'extensive museum' (*GR* 4, 450, 537).[37] But they were neither the first nor by any means the last to draw on those reserves. In 1938 Orson Welles's radio dramatization of *The War of the Worlds* revealed the ways in which the intersections of economic and social conditions, media codings, technical change, and historical events constituted what Hadley Cantril termed 'a pattern of circumstances providing a matrix for high suggestibility' – and therefore exploitation, however exciting.[38] Between 1950 and 1957 no less than 133 science fiction films – beginning, predictably, with *The Flying Saucer* – tapped the same vein. They used an expanding stockpile of mythological apparatus to dramatize contemporary existence as an extreme extension of established frontier myth. In the form of endless routines of panic and relief, invasion and dispersal, collapse and recovery,

---

36. Susan Sontag, 'The Imagination of Disaster,' in *Against Interpretation* (London: Eyre & Spottiswoode, 1967), pp. 209-25; Warren, pp. 58-9. Interestingly, given Clayton Chiclitz's plans to export refugee children to Hollywood to star in 'real big numbers, religious scenes, orgy scenes' (*GR* 558-9). Paramount originally purchased the rights to the novel (Philip Wylie and Edwin Balmer, *When Worlds Collide*) in 1932 as a possible vehicle for Cecil B. De Mille. Its provisional title was 'The End of the World.' Producer George Pal had worked at the UFA studios in Berlin from 1931 to 1932 and emigrated to the US in 1939. See Warren, pp. 3-4, 59.

37. William Burroughs, *Nova Express* (1966; London: Panther/Granada, 1968), p. 66. For the context see Vance Packard, *The Hidden Persuaders* (1957; Harmondsworth: Penguin, 1962) p.12.

38. Hadley Cantril, *The Invasion From Mars: A Study in The Psychology of Panic* (1940; New York: Harper & Row, 1966), pp. vi-vii.

they rehearsed the nightmare of defeat and destruction or vampiric absorption – personal, national, or global – and the dream of rejuvenating violence for security which the post-war state both encouraged and needed to constrain.[39]

But the enduring popularity and profitability of such films rested on more than power elite politics. Dreams of security within and liberation from imminent disaster are required, produced, and serviced inside a field of relationships and transformations, not simply imposed by authority. Following 1945, Cantril's 'pattern of circumstances' underwent an accelerating process of involution as the technological, economic, and social consequences of the war inflected American life, providing a 'matrix' of permanently induced, constantly changing, and continuously reinforced tensions. Technically, the stockpiles of science fiction barely outdistanced the experimental discoveries of scientists and their various expressions: during the final year of World War II the prosecution of the V–2 campaign against Antwerp, London, and Paris, and the destruction of Hiroshima and Nagasaki inaugurated a period of sustained intrusion.

In 1946, the world's first electronic digital computer was built at the University of Pennsylvania, RCA demonstrated a rudimentary color television system, the RAND Corporation opened in California, an engineer at Ford coined the term 'automation,' the Strategic Air Command was established, and the first post-war atomic bomb test took place at Bikini atoll. The following year, the Atomic Energy Commission began operation, the world's largest reflecting telescope at Mount Palomar was completed, the sound barrier was broken, the Zoomar lens was introduced (to cover baseball), and the first operational nuclear target list was compiled by the Joint Chiefs of Staff.[40] Between

---

39. Warren, pp. 442-5. See also Brian Ash (ed.), *The Visual Encyclopedia of Science Fiction* (London: Pan, 1977), pp. 294-5; Peter Biskind, *Seeing is Believing* (1983; London: Pluto, 1984), pp. 102-59; Nora Sayre, *Running Time: Films for the Cold War* (New York: Dial Press, 1982), pp. 191-202.

40. See Melvin Kranzberg and Carroll W. Pursell (eds), *Technology in Western Civilization. Volume II* (New York: Oxford University Press, 1967), pp. 320, 635; Erik Bernouw, *Tube of Plenty: The Evolution of American Television*, rev. edn (New York: Oxford University Press, 1982), pp. 101-2; S. Handel, *The Electronic Revolution* (Harmondsworth: Penguin, 1967), p. 155; David Alan Rosenberg, 'The Origins of Overkill: Nuclear Weapons and American Strategy, 1945-1960,' *International Security* 7, no. 4 (1983), pp. 12, 19; Gregg Herken, *The Winning Weapon: The Atomic Bomb in the Cold War, 1945-1950* (New York: Alfred A. Knopf, 1980), p. 213; R. Buckminster Fuller, *Critical Path* (1981; London: Hutchinson, 1983), p. 373.

1948 and 1951, IBM began marketing its first commercial electronic computer, television became a mass medium, the transistor was developed at Bell Laboratories (and soon after began displacing human operators in telephone exchanges), motivational research entered the armory of business operations, the US Army's Long Range Proving Ground at Cape Canaveral was established, and the term 'brainwashing' was first used in reference to communist re-education of Chinese citizens following the Nationalist defeat. During this same period, B.F. Skinner published *Walden Two*, Norbert Wiener issued *Cybernetics*, George Orwell introduced Big Brother and the Ministry of Love in *Nineteen Eighty-four*, and Isaac Asimov put together *I, Robot*. In addition, the Yale Technology Project began its study of car assembly plants and their workers, the US Army's 'Sandstone' test series demonstrated that atomic bombs could be produced along the same lines, the Soviet Union detonated its first atomic device, and President Truman ordered the production of an American hydrogen weapon. Finally, the first all-jet battle took place over Korea, the National Science Foundation was established, and the world's first experimental atomic breeder reactor commenced production of electricity in Idaho.[41]

Characteristically, since the 'pattern of circumstances' was traced and retraced within the framework of late capitalism and the Cold War, the development and utilization of science and technology took place unevenly. Whilst Vannevar Bush's *Science: the Endless Frontier* – begun in response to a request from President Roosevelt towards the end of World War II – advocated large-scale state support for post-war scientific research and development and thereby laid the groundwork for the National Science Foundation, it also fostered the expansion of a state-capitalist power complex centered on the armed forces, the armaments industries, and their scientists and technicians – the latest official expression of the Operation and its agents.[42] Provisioned by the terms of NSC-68 in April 1950, and by the outbreak

41. See Gilbert, pp. 31, 177-8; Garth Jowett, *Film: the Democratic Art* (Boston: Little, Brown & Co., 1976), p. 347; Packard, p. 28; John Marks, *The Search For The Manchurian Candidate* (London: Allen Lane, 1978), p. 125; Kranzberg and Pursell, pp. 93-4; Rosenberg, p. 19; Herken, p. 288; Samuel F. Wells, Jr., 'The Origins of Massive Retaliation,' *Political Science Quarterly*, 96, no. 1 (1981), p. 48; David Alan Rosenberg, 'American Atomic Strategy and the Hydrogen Bomb Decision,' *Journal of American History* 66, no. 1 (1979), pp. 62-87; *Lockheed Horizons* [house journal] 12 (1983), p. 38.

of the Korean War two months later, Bush's frontier was pushed through the following decade in directions which guaranteed a safe return – for some at least – investing additional American resources in the doctrine of national security, the fetish of preparedness, the survival of 'communism,' and the continued discovery of 'communists.'

Between 1952 and 1954, the Boeing B-52 moved from prototype to production model flight tests, North American produced the world's first operational supersonic fighter plane (the F–100 Supersabre), and the US Navy launched the world's first nuclear-powered submarine (the *Nautilus*). In Korea – primarily a Lockheed war – the Far Eastern Air Force celebrated its first unseen kill in aerial warfare, exhausted its 'air pressure' targets on the ground, and began attacking North Korea's rice irrigation system. Elsewhere, the Lockheed X–7 ramjet was successfully tested (and became the basis of the company's new Missile Systems Division), the Eniwetok hydrogen device (built for the renamed Defense Department by Du Pont) was exploded, and atomic artillery was developed by the US Army. Furthermore, the US Air Force's long-range rocket study group – chaired at the Massachusetts Institute of Technology by John von Neumann – was charged with investigating the feasibility of an intercontinental ballistic missile. Their report found such weapons not only feasible but unstoppable, apocalyptic, and therefore essential – within two years American investment in rocketry began to mushroom.[43] Between 1955 and 1957, the Technological Capabilities Panel Report to President Eisenhower recommended the

42. See James R. Killian, Jr., *Sputnik, Scientists, and Eisenhower* (1977; Cambridge, Mass.: MIT Press, 1982), p. 58; Vannevar Bush, *Modern Arms and Free Men* (London: Heinemann, 1950), pp. 278-91; Michael S. Sherry, *Preparing For The Next War: American Plans for Postwar Defense, 1941-1945* (New Haven and London: Yale University Press, 1977), pp. 120-58; Yergin, pp. 266-9; C. Wright Mills, *The Power Elite* (1956; New York: Oxford University Press, 1959), pp. 216-19; Jürgen Habermas, 'Problems of Legitimation in Late Capitalism,' in Paul Connerton (ed.), *Critical Sociology* (Harmondsworth: Penguin, 1976), pp. 365-9.

43. See Boeing Inc., *Background Information: The Boeing Company* (Seattle: Boeing, 1982), p. 7; Christopher Chant *et al.*, *The Encyclopedia of Air Warfare* (London: Spring Books, 1985), p. 191; Sampson, pp. 99-100; Callum MacDonald, *Korea: The War Before Vietnam* (London: Macmillan, 1986), pp. 175, 234-41; *Lockheed Horizons* 12 (1983), p. 38; Rosenberg, 'Origins of Overkill,' pp. 30-45; Norman Moss, *Men Who Play God: The Story of the Hydrogen Bomb* (1968; Harmondsworth: Penguin, 1970), p. 209; Dwight D. Eisenhower, *The White House Years: Waging Peace, 1957-1961* (London: Heinemann, 1966), p. 208; Schoettle in Lakoff, pp. 165-8.

highest national priority for ICBM and IRBM construction, the dispersal of SAC bases, and the initiation of anti-ballistic missile research; the first B–52 bombers were delivered to SAC, along with hydrogen bombs from the AEC and an IBM–704 computer for target data analysis; U–2 high altitude intelligence-gathering overflights of the Soviet Union started, and the DEWLINE system was constructed and activated (both as recommended by the TCP Report); and the Gaither Committee indicated further investment opportunities by proposing an expansion of conventional military resources and a $25-billion fallout shelter program.[44]

But if Clayton Chiclitz's outlying and potentially painful network was flourishing (and in the process dealing Franz Pökler back into a new 'game' with much higher stakes), opportunities were also being invented, discovered, and exploited in the ostensibly happier domestic settlements, thus projecting Gerhardt von Göll's dream beyond the celluloid and bringing his 'seeds of reality' to fruition. The success of the 'Bring Back Daddy' campaigns and the failure of universal military training legislation following World War II only bolstered voluntary mobilization for the recovery in civilian life of what the war had simultaneously justified, fostered, and postponed. When, in 1947, cinema box office receipts fell off for the first time in a decade, the opening of Congress was televised, and the first (whites only) Levittown was settled on Long Island, national security extended its theater of operations into a new and more dispersed set of installations.[45]

The extension had its costs. Between 1945 and 1960 some $116 billion was borrowed from financial institutions to cover house mortgages for the twenty-three million people either born, marrying, or moving into a suburbia that was itself growing around forty times as fast as central city areas. But the customers progressively took on the load. Within twenty years of the war's end, total short-term consumer debt rose from $5 billion to $74

---

44. See Killian, pp. 67-87, 96-101; Rosenberg, 'Origins of Overkill,' pp. 37, 45, 47-8.

45. See Gaddis, pp. 261-2, 341; Sherry, pp. 73-83; Maltby, p. 63; Gilbert, pp. 115-16; Bernouw, p. 101. On this dispersal see William H. Whyte, *The Organization Man* (1956: Harmondsworth: Penguin, 1960), pp. 246-361. Whyte's picture obviously gives a stereotype – not necessarily inaccurate, but tendentious. Herbert Gans, *The Levittowners: Ways of Life and Politics in a New Suburban Community* (New York: Pantheon, 1967) provides depth and qualifications.

billion to finance not just the 'pin-ups and library shelves' of the Orpheus Theater's 'invisible rooms' but the entire consumer goods fallout blanketing the new private quarters.[46] Two commodities carried most weight: between 1947 and 1957 automobile credit multiplied eightfold – in 1957 65 per cent of all new cars were bought by installments; between 1947 and 1960 the number of television sets in America increased from 50 thousand to 50 million – during the mid-50s 10 thousand new sets were installed every day. American life was being reset and reprogrammed rapidly. In 1951 *See it Now* replaced *Hear it Now* and more television sets were sold than radios. In 1952 *TV Guide* first appeared, as did Walter Cronkite – the first anchorman – and J. Fred Muggs, a chimpanzee whose presence boosted the ratings for the *Today* program. While Adlai Stevenson refused to be sold 'like a breakfast food,' 'I like Ike' proved a popular jingle and propelled the former Supreme Commander into the White House. In 1953 the coronation of Queen Elizabeth II stimulated sales of receivers, while the birth of Lucille Ball's son on *I Love Lucy* attracted more viewers than Eisenhower's inaugural address. Tastes proved fickle. By 1954 Americans were fully prepared for the TV dinner – and televised *Disneyland*.[47]

But the box in the corner radiated more than its scheduled output: even as millions of 'our boys' were retained by the armed services under an extended draft law to help oversee the rehabilitation of Europe and Asia, television was being enlisted as both vanguard and sentry for a deepening occupation at home. In 1946 advertising agencies were already experimenting with

---

46. This extension was predicated on what Aglietta terms the 'intensive regime' of accumulation ushered in by the technologies of Fordism and the politics of the New Deal. See Michel Aglietta, *A Theory of Capitalist Regulation: The US Experience*, trans. David Fernbach (London: Verso, 1979), pp. 116-17, 151-61; Mike Davis, *Prisoners of the American Dream* (London: Verso, 1986), pp. 190-91.

47. See Theodore H. White, *The Making of the President, 1960* (1961; London: Jonathan Cape, 1964), pp. 217-19, 279; Kranzburg and Pursell, p. 90; Vance Packard, *The Waste Makers* (London: Longmans, 1961), p. 151; Bureau of the Census, *Statistical Abstract of the United States: 1961* (Washington: US Government Printing Office, 1961), p. 560; Charles C. Alexander, *Holding the Line: The Eisenhower Era, 1952-1961* (Bloomington: Indiana University Press, 1975), p. 103; Robert Sobel, *The Manipulators: America in the Media Age* (New York: Doubleday/Anchor, 1976), p. 307; Jowett, p. 347; Bernouw, pp. 136-7, 145-8, 170; Gilbert, p. 66; Douglas Miller and Marion Nowak, *The Fifties: The Way We Really Were* (Garden City, N.Y.: Doubleday, 1977), p. 7; Richard Schickel, *The Disney Version*, rev. edn (London: Michael Joseph, 1986), p. 314.

commercials, and in 1947 Camel cigarettes and Oldsmobile agreed to sponsor NBC and CBS news programs. The narcotics industry was integrating horizontally: within ten years sponsorship would itself constitute a billion-dollar operation. During the 1950s television pushed the market home, distributing automobiles, cigarettes, headache tablets and stomach pills – *New Dope* could be shot, smoked, *and* swallowed – to fund its own extension. Vacuum cleaners, dishwashers, and washing machines combined to soak up domestic funds – in the process eliminating domestic servants – whilst the long-playing record (perfected by Goldmark) and the Polaroid camera (marketed by Land) provided other options from 1948 onwards and furnished additional outlets for the Imipolex still pouring from the Operation's tower reactor in *Gravity's Rainbow*. When the universal credit card was introduced by Diner's Club in 1950, its carriers were spoiled for choice (*GR* 486-8, 745).[48]

By 1956, when the United States became the world's first service economy and the first country in which the population spent more time watching television than working, American society seemed to be as high as it could get – a condition Sinatra and Crosby's 'swell party' at Newport punctually celebrated. With 81 per cent of families in possession of – or in the possession of – television sets, 89 per cent with washing machines, and 96 per cent with refrigerators, the elevator appeared overloaded and its integrated circuit of 'invisible rooms' saturated.[49] But E.L. Bernays' *The Engineering of Consent*, published the previous year, provided an answer to the limits of growth: 'the creation of dissatisfaction with the old and outmoded. The engineering of planned obsolescence via annual style changes, introduced by Alfred Sloan and Harley Earl at General Motors in the 1920s, was quickly adopted by domestic appliance manufacturers to help – in von Göll's words – 'keep it going indefinitely.' In 1956 style and color changes came to fridges, stoves, and typewriters; the same year the New York Telephone Company offered customers multicolored extensions allowing 'impulse phoning' to help 'eliminate the tension.'

48. See Gaddis, pp. 261-2; Sherry, pp. 191-4; Bernouw, pp. 100-101, 198-9; Fuller, pp. 387-8; Packard, *Waste Makers*, p. 150; Gilbert, p. 29.

49. For these statistics, see Victor Fuchs, 'The First Service Economy,' *Public Interest* 2 (Winter, 1966), pp. 7-17; Alexander, pp. 104, 113; Miller and Nowak, p. 7. On the consequences see Norman Mailer, 'From Surplus Value to The Mass-Media,' in *Advertisements for Myself* (New York: 1959; London: Panther/Granada, 1968), pp. 353-6.

The imperial organization men – denounced in Engelbrecht and Hanighen's *Merchants of Death* in 1934 but resurrected and refinanced by World War II – were being supplanted in the incipient totalitarian era by self-styled 'merchants of discontent.' In some cases – notably the Du Ponts – they were one and the same (*GR* 745).[50]

The New York Telephone Company offer – one of the '1,518 selling messages' that 'a typical American family is exposed to ... in the course of an average day' – may have tapped a prime market: between 1950 and 1960 the number of telephones in the United States increased from 43 million to 74 million while the daily number of calls rose from 175 million to 282 million. According to Marshall McLuhan, however, this growth helped eliminate only such things as the inviolability of management offices, not the 'tension' behind the unmade connection.[51] The latter regenerated – and could be repeatedly engaged – because the power structure remained intact and because the 'merchants of discontent' required precisely that cycle of recurring dissatisfaction. The Rocket State thrived and rested on tension. It constituted a network of interlocking contradictions: contained by dispersal, absorbed by commodities, accumulated by consumers, adjusted and resupplied by the media – but never resolved. And the meaning of this was revealed, not in the selling points of *High Society* – its universe of ease within security which marketed class power as natural order – but across the entire structure of production and consumption on which it relied. Not everyone could be Bing Crosby or Grace Kelly. In 1954 Samuel Stouffer's opinion survey, *Communism, Conformism, and Civil Liberties*, found Americans primarily worried about personal financial and health problems: loan repayments, overdemanding children, mortgages, college finance, hospital bills, and so on. Less than 1 per cent mentioned communism or civil liberties. By 1960 Packard's *The Waste Makers* included the more prevalent consequences of enclo-

---

50. See Packard, *Hidden Persuaders*, pp. 25, 144; Alfred P. Sloan, Jr., *My Years With General Motors* (1963; New York: MacFadden-Bartell, 1965), pp. 265-78; Helmuth C. Engelbrecht and Frank C. Hanighen, *Merchants of Death: A Study of The International Armaments Industry* (London: George Routledge & Sons, 1934), pp. 22-37, 180; Packard, *Waste Makers*, passim.

51. See Packard, *Waste Makers*, p. 216; Bureau of the Census, *Statistical Abstract of the United States: 1951* (Washington: US Government Printing Office, 1951), p. 453; *Statistical Abstract: 1961*, p. 508; Marshall McLuhan, *Understanding Media* (1964; London: Abacus, 1973), pp. 289-91.

sure by the multiplying forms of elevated living – what William Burroughs termed 'junk' – within a competitive, parasitic, and insecure society: anxiety, guilt, various family and medical problems, theft, even a kidnapping to raise funds for mounting credit charges.[52]

## Travel Arrangements

Now, plunged into a nighttime far deeper than that from which this morning they awoke (or thought they did), the people seek – with distraught hearts and agitated loins – a final connection, a kind of ultimate ingathering, a tribal implosion, that will either release them from this infinite darkness and doleful sorrow or obliterate them once and for all and end their misery.

— Robert Coover, *The Public Burning* (1977)

Have you heard – it's in the stars
Next July we collide with Mars!
Well, did you evah?
What a swell party this is!

— Frank Sinatra and Bing Crosby, 'Did You Evah?' *High Society*
(1956)

The 'action' of the Cold War – from the 'excitement' of *The War of the Worlds* to the scares over Strontium–90, from the thrill of an Oldsmobile to the threat of repossession – therefore proves no more than a set of exercises in stasis. It mobilizes the elevator, furnishes the 'invisible rooms,' and restocks the 'peep-show machines' of the Orpheus Theater, yet these moves do not 'eliminate the tension' but merely disperse, upgrade, and reproduce it across the Rocket State's sprawling joints. It accommodates, adjusts, and advances Clive Mossmoon and his progeny, but in the process leaves them besieged by anxiety and commitments. It engineers novelty simply as a means of extending necessity, and thrill as a means of defending routine. As a result it

---

52. Samuel Stouffer, *Communism, Conformism, and Civil Liberties: A Cross-section of the Nation Speaks Its Mind* (1955; New York: John Wiley & Sons, 1966), pp. 59-60, 68; Burroughs, *Naked Lunch*, pp. 8-10; Packard, *Waste Makers*, pp. 156-7, 235, 241; Vance Packard, *The Status Seekers* (1959; Harmondsworth: Penguin, 1961), pp. 223-31.

provides not relief but progressive entanglement in an indistinct region of undeclared combat between the proliferating forms and forces of 'Happyville' and 'Pain City': domesticated ease and colonial tensions, open plan suburbs and armored subways, shopping centers and insane asylums, upholstered lounges and padded cells. It reproduces and consumes post-war America at the precarious interface of inflicted exterior power and imagined interior collapse, of sanctioned enforcement and feared retribution.

But this diffuse struggle takes on specific form as a succession of set pieces within an immanent propaganda war. Combat is joined, campaigns waged, and irregulars conscripted around the banner of the *New York Times* – the conditions are neatly encapsulated in Robert Coover's *The Public Burning* – and the other propagative agencies of the Rocket State. Thus in 1955, one of the 122 nuclear weapons tests carried out by the United States between 1951 and 1958 – which collectively constituted the central act in Washington's rehearsals for security – was set up by the government as a public relations exercise designed to give the man in the street (or the lounge) a sense of their power. In 'Operation Cue' some 200 companies decorated the blast area with their various consumer products in order to test their durability and reliability under demanding conditions while television commentators and press men crouched in a trench two miles from Ground Zero waiting to publicize the results. Six years later *Life* magazine informed its readers that over 90 per cent of the population could survive nuclear attack in fallout shelters. The following year the American Institute of Architects announced a national competition to find the 'best blast-proof building of the year.' Reinforcement of the elevator and its network of invisible rooms might, it seemed, be sufficient.[53]

But behind these screens of reassurance, images of the nightmarish and grotesque also thrived. In October 1949, after the detonation of a Soviet atomic bomb was announced by President Truman, the *New York Journal-American* printed a half column picture showing Manhattan engulfed in atomic 'waves of

53. Ellul, *Propaganda*, pp. 144-7; Robert Coover, *The Public Burning* (1977; Harmondsworth: Penguin, 1978), pp. 237-48, 265-72, 351-7, 595-607; Miller and Nowak, p. 56; Gilbert, pp. 172, 174; Theodore Sorensen, *Kennedy* (London: Hodder & Stoughton, 1965), p. 615. For 'Operation Cue,' see *Life*, 16 May 1955, p. 58, 30 May 1955, pp. 39-42.

havoc and death,' whilst the *Chicago Tribune* reminded its readers that its publisher had already built himself a fallout shelter. In March 1951, as American forces pushed north against Chinese troops in Korea, a *New York Times* headline announced 'DANGER OF ATOM BOMB ATTACK IS GREATEST IN PERIOD UP TO THIS FALL,' whilst the *New York Journal-American* repeated that New Yorkers would be the first to see 'action.' In September 1952, as the rhetoric of 'rollback' rolled Eisenhower on towards the White House, *Life* circulated a photo-essay on disfigured survivors of Hiroshima and Nagasaki. And in May 1955, following the publication of the disturbing Atomic Energy Commission report on fallout, the *US News and World Report* asked 'What Will Radiation Do To Our Children?' Would the elevator be deformed into a 'moving wood scaffold' after all?[54]

Imaginative answers had already begun clambering across the nation's screens with the release of *The Beast From 20,000 Fathoms* in 1953 – Manhattan yet again bears the brunt – and continued to surface for much of the decade. And whilst the terms of a final solution remained moot until 1962, when the American Institute of Architects announced its first prize winner two months *after* the Cuban missile crisis, other conditional reactions built up impulsively across the social structure. In 1953, as the Cold War moved towards the climactic and sacrificial execution of the Rosenbergs, the answer of millions of Robert Coover's *New York Times* 'communicants' to the prevailing 'pattern of circumstances' – '*Let me Out!*' – was both hallucinatory and exemplary.

As Jacques Ellul notes in *Propaganda*, those 'subjected to two intense, opposing propagandas' – the 'contradictory commands' which according to William Burroughs are 'the basic formulae of the daily press' and 'an integral part of the modern industrialized environment' – seek escape, characteristically via the distinct but intersecting methods of inertia and involvement. They attempt to flee the contradictions of the Rocket State through resignation to the apolitical or through commitment to the partisan; they evade 'the opposing clash of propagandas' by passively retreating into privacy or by actively aligning within the polity. In *The Public Burning* the 'Friday-morning commuters' shuttle rapidly between the two: at one point 'beset with nightmare visions of Soviet tanks in Berlin, dead brothers lying scattered across the cold wastes of

---

54. Joseph Goulden, *The Best Years, 1945-1950* (New York: Atheneum, 1976), pp. 246-7; Coover, p. 34; Miller and Nowak, p. 56; Moss, pp. 110-11.

Korea, spreading pornography and creeping socialism, Phantom-ized black and yellow people rising up in Africa and Asia,' and at the next 'dreaming peacefully of baseball, business, and burning hayricks' under the joint command structure of Uncle Sam, Nelson Eddy, and Frankie Laine.[55]

The movement of Coover's cross-section of the lonely crowd, his image of Mossmoon's manifold progeny, describes the Friday-morning commuters' domain as the deepening interface between Ellul's complementary techniques of evacuation: those points where notional withdrawal to the 'apolitical' sphere of bourgeois privacy constitutes *de facto* commitment to the partisan, and where partisan alignment in turn advances effective depoliti-cization, since each confirms the extension of civil society within capitalism. But the commutation drives both inertia and involve-ment beyond the short circuits of bourgeois politics. For if, as Ellul notes, an evacuee's resignation from the latter 'progressively takes over the whole of his being and leads to a general attitude of surrender,' then equally his 'flight into involvement' reaches well beyond – and may in the process obviate – his commitments to mere party or leader. The dream of escape obeys and violates whatever is necessary.[56]

In both cases public reaction to Welles's broadcast of *The War of the Worlds* forecast things to come. Through his interviews Hadley Cantril discovered that for thousands of listeners embedded in routine lives the attendant 'psychology of panic' included a variety of liberatory and cataclysmic fantasies: dreams of escaping from paying bills and saving money, of freedom from the police (and mothers-in-law), and a pervasive thrill at being vicariously involved in global disaster. At the end of the 1950s Leslie Fiedler summarized the science fiction genre as an extended tapping of the audience's 'masochistic delight in imagining a future in which mutants, robots, extra-terrestrials, dogs, or simply *nothing* takes over.' Six years later, in 1965, Susan Sontag concurred that the previous decade's output enacted a spectacular fantasy of personal survival inside universal destruc-

---

55. Warren, pp. 99-104; Coover, pp. 136-9, 247; Ellul, *Propaganda*, pp. 181-2; William Burroughs (with Daniel Odier), *The Job* (London: Jonathan Cape, 1970), pp. 30-31.

56. See Jacques Ellul, *The Technological Society*, trans. John Wilkinson (New York: Vintage, 1964), p. 371; Marx, *Grundrisse*, p. 156; Ellul, *Propaganda*, p. 182; Wilson, pp. 174-5.

tion which allowed each viewer to 'participate in the fantasy of living through [his or her] own death and ... the destruction of humanity itself.'[57] And if these conclusions were synoptic, they did attest to the existence of a widespread dream of 'shelter in time of disaster' via inertia – a fantasy variously indicated by the mass popularity of J.D. Salinger's *The Catcher in the Rye* (1951), of organized religion and President Eisenhower, of newer opiates such as the tranquillizers Miltown and Thorazine (sales of which were valued at almost $5 million annually within five years of their introduction in 1954), and of sleeping pills. The definitive articulation of this fantasy was Don Siegel's *Invasion of the Body Snatchers*, released in 1956, the same year as *High Society*.[58]

On the other hand, Cantril's research into the repercussions of Welles's radio play also concluded that 'the extreme behavior evoked by the broadcast was due to the enormous felt ego-involvement the situation created and to the complete inability of the individual to alleviate or control the consequences of the invasion.' Likewise, Leslie Fiedler found science fiction stories of the 1950s pandering to 'dreams of omnipotence' within the experience of powerlessness, whilst Susan Sontag detected in the film versions an insatiable 'hunger for a "good war" which poses no moral problems [and] admits of no moral qualification.' If the fantasy of generalized disaster vicariously released those tuned into *The War of the Worlds* from routine obligations, it also provided their descendants with 'a fantasy target for righteous bellicosity to discharge itself.' In the wake of Hiroshima and the redeployment of imperial power, the confrontation between alien predatory mutant and vulnerable innocent human effortlessly sanctioned a violence henceforth officially deplored for organizational and technical reasons: Things, zombies, and blobs progressively supplanted 'Indians,' 'Krauts,' and 'Gooks' within the unquestioned recognition patterns of a society attuned to active service in war and peace.[59]

But if these evasive actions share a common stimulus, the

57. Cantril, pp. 161-2; Leslie Fiedler, *Love and Death in the American Novel*, 2nd edn (London: Jonathan Cape, 1967), pp. 501-2. Cf. Thomas Pynchon's own memoir in his *Slow Learner: Early Stories* (London: Jonathan Cape, 1985), p. 13.

58. Miller and Nowak, pp. 84-99, 138; Gilbert, pp. 238-40; Mailer, *Presidential Papers*, p. 57. On Siegel's *Invasion of the Body Snatchers*, see Warren, pp. 281-90; Stuart Samuels, 'The Age of Conspiracy and Conformity: *Invasion of the Body Snatchers*,' in John E. O'Connor and Martin Jackson (eds), *American History/American Film* (New York: F. Ungar, 1979), pp. 203-17.

social responses they induce may diverge. Whereas 'refuge in inertia' constitutes an acceptable installation within the Rocket State's circuit of 'invisible rooms,' 'flight into involvement' threatens to antagonize both neighbors and house management. When Holden Caulfield plans to become an anonymous, deaf mute filling station attendant in *The Catcher in the Rye*, he moves towards an anarchic isolationism which, whilst stoic, is also conservative, futile, and rife for accommodation: in the war after the war anyone can take a surrender. But when 'righteous bellicosity' expresses itself at places like Little Rock, Kansas, it strains the elevator, slows its ascent, and thereby rattles the Operation in its efforts to tempt in and integrate more fans. The latter's ideal enforcement agency remains epitomized by Mickey Spillane's fictional detective, Mike Hammer. In a series of best-selling novels beginning with the aptly titled *I, the Jury* (1947) and *Vengeance Is Mine* (1950), Hammer expresses the secret dream of a society committed to limitless competition and personal deliverance: the private deployment of strength against formulaic 'evil' in the interests of implicit social recovery without radical social change; the righteous conquest of intrusive 'corruption' and 'lawlessness' in defence of reaffirmed personal and communal innocence. Through formal routines of detection and exposure which barely conceal the interior drives, Hammer embodies the characteristic compensatory super-self of those politically disarmed and socially insecure would-be *Führers* who find themselves enclosed by the unexplained restrictions of an unacknowledged system: the self-satisfying vigilante of incipient totalitarianism, the articulating unit of Fortress America.[60]

But the myth user takes up only what he wants to need, and since Coover's commuters move at the interface of obedience and anarchistic power, the no man's land's increasingly absorbing joint, their social codes adjust accordingly. Norman Vincent Peale may have institutionalized himself in the media – along

---

59. Cantril, p. 200; Fiedler, p. 501; Sontag, pp. 215, 218; Biskind, p. 164. See Les Daniels, *Fear: A History of Horror in the Mass Media* (St. Albans: Panther/ Granada, 1977), pp. 176, 183-204; Eric Mottram, 'Out of Sight But Never Out of Mind: Fears of Invasion in American Culture,' *Talus* 1 (Spring, 1987), pp. 43-79.

60. J.D. Salinger, *The Catcher in the Rye* (London: Hamish Hamilton, 1951), pp. 236-7; M. Thomas Inge (ed.), *Concise Histories of American Popular Culture* (Westport, Conn.: Greenwood Press, 1982), p. 114; Miller and Nowak, p. 168. For an outline of the conditions for Spillane's popularity, see Alan Harrington, *Psychopaths* (London: If Books, 1972), pp. 17-48.

with Billy Graham and Fulton Sheen – by preaching and profiting from a formula of personal adjustment, devotion to 'God,' and faith precluding social analysis, but *The Power of Positive Thinking* – the title is again indicative – topped the best-seller lists until 1955 by also offering to 'release your inner power' in the interests of 'boundless energy' and personal 'strength.' Mike Hammer may have imposed salvation in a more overtly muscular fashion, but his permission as enforcer rested on his own necessary submission: beginning as an innocent voyeur – the private eye – within the corrupt city, he gains sanction for vengeance only when subjected to the enemy's villainous arsenal of guns, drugs, and enticements – things the reader would not, of course, touch.[61]

Spillane's allegory of power is as false as those rehearsed by his precursors and allies, from Dick Tracy to Superman, from James Bond to Buck Rogers. His society is not a community but a privatized sacrificial combat–state patrolled between the uneasy sanctuaries of home and work under conditions more accurately recorded by Hubert Selby in *The Demon*.[62] But for the descendants of General Cummings such fantasies exerted their power. Likewise, Peale's combination of self-reliance, adjustment, and pre-New Deal conservatism (he opposed income tax, minimum wage legislation, and the welfare state, considered Roosevelt a dictator and so naturally supported MacArthur for the Presidency in 1948) is a cruelly appealing fraud for millions more trained to believe in his Horatio Alger-style parables of men like Cummings. Since their thrilling or calming routines are only part of the continuing 'action' on display in the post-war theater, they solve nothing and become endlessly repeatable stock features. Since the customer always needs another shot, the junkie another vein, their desires are perpetuated by spurious novelty: Burroughs' 'Trak servicing.' The 'revue' frames them ceaselessly. In 1957 over 100 television series hit the market with titles which interlocked easily: *Highway Patrol, Trackdown, Lone Ranger,*

61. Miller and Nowak, pp. 84-99; Gilbert, pp. 238-40; Norman Vincent Peale, *The Power of Positive Thinking* (1952; Surrey: World's Work, 1953), pp. 1, 41; Fiedler, pp. 347-8.

62. Arthur Asa Berger, *The Comic-Stripped American* (New York: Walker & Co., 1973), pp. 93-101, 122-32, 146-58; Marshall McLuhan, *The Mechanical Bride: Forklore of Industrial Man* (1951; Boston: Beacon Press, 1967), pp. 102-3; Eric Mottram, 'Living Mythically: The Thirties,' *Journal of American Studies* 6, no. 3 (1972), pp. 276-9; Hubert Selby, *The Demon* (1976; London: Marion Boyars, 1977).

*Official Detective, Superman, Frontier, Border Patrol,* and so on. Eisen-hower turned on for *Wild Bill Hickock.* Advertisers slotted in with commodities which absorbed the heroes' magical powers of salvation. Sponsors licensed violence since it sped the hidden hand. *Perry Mason* aside, constitutional structures of justice were clearly insufficient: for America in the 1950s the dream of security remained paramount.[63]

## Watch the Skies

> She flips a red lever on her intercom. Far away a buzzer goes off. 'Security.' Her voice is perfectly hard, the word still clap-echoing in the air as in through the screen door of the Quonset office with a smell of tide flats come the coppers, looking grim. Security. Her magic word, her spell against demons.
>
> — Thomas Pynchon, *Gravity's Rainbow* (1973)

The propagative center for the newly developed doctrine and 'state of mind' of national security remained the power elite, who from 1947 onwards accelerated their investments in 'commun-ism' – the nation's second 'Red Bank' – and required their supporting cast to do likewise. Early that year former US Chamber of Commerce President Eric Johnston brought the State Department's message to Hollywood:

> the rifle the film industry had shouldered in World War II could not be put down; it had to keep marching to the drums of another martial conflict – the Cold War with international communism.

The results in terms of output have been thoroughly documented and analyzed. If films like *The Iron Curtain* (1948) and *My Son John* (1952) were little more than cheaply made and rarely profit-able loyalty tests for Hollywood, they were also widely screened as second features – the undercarriage to *High Society* – and constituted effective branch agencies for the dispersal and circul-

---

63. Miller and Nowak, p. 105; William Burroughs, *The Soft Machine* (New York: Grove Press, 1967), pp. 46-7; Bernouw, pp. 132, 142, 213-16, 261-5, 405-7. 'Fair Play' with Communism had in fact been written off as early as 1947 accord-ing to Herbert Hoover's secret study on covert operations, quoted in Marks, p. 27.

ation of the image bank's latest legal tender.[64]

Reinvestment came not only in movies from other genres such as *Walk East on Beacon*, 'produced with the cooperation of the FBI' in 1952, and *Invasion USA*, an hallucinatory 1953 science fiction fantasy of American defeat in the next war, but right across the media: in Mickey Spillane's *One Lonely Night* (1951); in television series like *Man Against Crime* and *I Led Three Lives*; even in advertising, where 'Fidelity, Bravery, and Integrity' were qualities superimposed on the varied products of companies from Ford Business Insurance to Frozen Beans Incorporated – much to the annoyance of the FBI. Moreover, the bank's other branches regularly corresponded: Gordon Douglas's *I Was a Communist for the FBI* (1951) fulfilled Jacques Ellul's requirements for effective propaganda on its own by appearing as a feature film (nominated for an academy award as best documentary), television series, *Saturday Evening Post* trilogy, bestselling novel, radio serial, and lecture tour. In addition, their currency was fully convertible, so that Gordon Douglas could go on to make *Them!* in 1954 as a hard-boiled crime thriller in disguise, whilst *The Red Menace* was released 'with unusual pride' in 1949 by Republic, a studio which had previously specialized in monster movies and westerns.[65] Finally, exchange came in from every corner. Television may have constituted the decade's most effective fifth column – and one which Arch Oboler had laborious fun with in *The Twonky* (1953) – but the banking system also transferred the credit for brainwashing to the Soviets' inexhaustible account following the initial depositions in the 1949 trial of

---

64. Yergin, pp. 193-201; Sherry, p. 133; Peter Irons, 'American Business and the Origins of McCarthyism: The Cold War Crusade of the United States Chamber of Commerce,' in Robert Griffith and Athan Theoharis (eds), *The Specter: Original Essays on the Cold War and the Origins of McCarthyism* (New York: New Viewpoints, 1974), pp. 72-89; I.F. Stone, *The Truman Era* (1953; New York: Vintage, 1973), pp. 80-86; Eric Mottram, 'The Red Bank: Indians in White Lives,' in *Essays by Eric Mottram and Philip Davies* (London: Polytechnic of Central London, 1978), pp. 17-37; Ceplair and Englund, p. 249. On Hollywood's response see the sources listed on p. 86 n. 23, and also Robert Sklar, *Movie-Made America: A Cultural History of American Movies* (1975; London: Chappell & Co., 1978), pp. 256-66; Sayre, pp. 18, 22, 79-81.

65. Sayre, pp. 31, 84, 86, 91; Fiedler, p. 502; Miller and Nowak, p. 29; Bernouw, pp. 130-34, 217; David Zane Mairowitz, *The Radical Soap Opera* (1974; Harmondsworth: Penguin, 1976), p. 126; Lawrence Alloway, *Violent America: The Movies, 1946-1964* (New York: Museum of Modern Art, 1971), p. 42; Ellul, *Propaganda*, pp. 9, 17-18; Brosnan, p. 98; Colin Schindler, *Hollywood Goes to War* (London: Routledge & Kegan Paul, 1979), pp. 120-21.

Cardinal Mindszenty. As Stouffer's 1954 survey revealed, hardly any American could claim even to have met a 'communist.' They constituted a flexible foe which could be turned into anything and drawn on at will.[66]

Propagandists from the American Legion to HUAC, from Adlai Stevenson to J. Edgar Hoover, sought access quickly. Joseph McCarthy drew heavily on the account of his friend and fund raiser John Wayne: to citizens totally familiar with routines of hoodlums, mad scientists, and saboteurs whose evil required exposure, his warnings that 'one communist with a razor blade poised over the jugular vein of this nation, or in an atomic energy plant, can mean the death of America,' played the Senator's negligible discoveries into substantive mythologies. But in the incipient totalitarian state such routines are self-validating. By 1955, when *Strategic Air Command* picked up where *The Thing From Another World* (1951) and Harry Truman left off, entertainment, news, and public affairs were being deposited in a barely differentiated image bank whose sociopolitical history informed the activities of Americans across the class structure. Between the predictions of the National Security Council's adjunct to the latest Joint Chiefs of Staff war plan DROPSHOT, drafted in early 1949, and the dramatization of such activities in William Cameron Menzies' *The Whip Hand*, first screened in October 1951, there was already little distance. Gerhardt von Göll's 'seeds of reality' had fallen on fertile ground.[67]

The extent of that fertility was confirmed by the early scenario for George Pal's *When Worlds Collide*, released a month after *The Whip Hand* in November 1951. Its classification of potential

66. Warren, pp. 149-51; Brosnan, p. 100; Sayre, pp. 25-6, 80-82, 201-2; Donald F. Crosby, 'The Politics of Religion: American Catholics and the Anti-Communist Impulse,' in Griffith and Theoharis (eds), pp. 24-6; Marks, pp. 125-31; Stouffer, pp. 175-6.

67. See Goulden, p. 308; Sayre, pp. 13-14, 21; Coover, pp. 134-5; Athan Theoharis, 'The Rhetoric of Politics: Foreign Policy, Internal Security, and Domestic Politics in the Truman Era, 1945-1950,' in Barton J. Bernstein (ed.), *Politics and Policies of the Truman Administration* (Chicago: Quadrangle, 1970), p. 215; Thomas C. Reeves, *The Life and Times of Joe McCarthy* (London: Blond & Briggs, 1982), pp. 326, 430, 545; Adler in Griffiths and Theoharis (eds), p. 347; Brosnan, pp. 84-8; Biskind, pp. 64-9; Athan Theoharis, *Seeds of Repression: Harry S. Truman and the Origins of McCarthyism* (Chicago: Quadrangle, 1971), pp. 63-4; Julian Smith, *Looking Away: Hollywood and Vietnam* (New York: Charles Scribner's Sons, 1975), pp. 190-93; Sobel, p. 328; Alloway, p. 30; Herken, pp. 284-5; Warren, pp. 66-70.

public behavior had foreseen 'some in religious resignation.' On publication the following year, the Revised Standard Version of the Bible sold two million copies and challenged Peale's *The Power of Positive Thinking* at the top of the best-seller lists. The film's scenario envisaged 'some in terror,' a prediction borne out by the enormous success of *King Kong* (1933) on its re-release in the summer of 1952, and by the popularity of the mutant boom it spawned. It anticipated many more 'in a terrific orgy of rape, lust, robbery, murder, etc.,' occupations whose attractions could be quickly gauged by a check of the top rating television programs, the contents of comic books and comic strips ('the funnies'), and the popularity of Mickey Spillane's novels, the Kinsey Reports of 1948 and 1953, and Hugh Hefner's *Playboy*, launched in December 1953 with the assistance of Marilyn Monroe. Finally, the scenario included amongst its propositions 'some struggling by science to escape.' And it was this method – adopted successfully in *When Worlds Collide* – which offered the most secure of all possible ways out. Moreover, it was not only George Pal's randomly selected celluloid survivors who stood to gain. The specific mechanism of evacuation – the 'huge rocket' proposed in the scenario – also promised a solution for a more willfully induced alliance of volunteers and conscripts: a cross-section of customers, engineers, organizers, and leaders which, even as the earth was being dashed to pieces on film, was starting to install itself on all sides of the screen and getting ready – like the fans in the Orpheus Theater – for the real show to begin.[68]

Production had in fact already begun, for *When Worlds Collide* capitalized partly on the success of George Pal's previous film, *Destination Moon*, and partly on the impact of its historical sources. Released in 1950, *Destination Moon* was – with the exception of Fritz Lang's pioneering *Die Frau im Mond* (1928) – the first science fiction film to portray space travel as a practical option. After its profitability had encouraged Paramount to hire Pal as the producer of *When Worlds Collide*, its critical reception

---

68. See Warren, pp. xiv, 59; Miller and Nowak, p. 86; Otto N. Larsen (ed.), *Violence and the Mass Media* (New York: Harper & Row, 1968); Gershon Legman, *Love and Death: A Study in Censorship* (1949; New York: Hacker Art Books, 1963); David Manning White and Robert Abel (eds), *The Funnies: An American Idiom* (New York: Macmillan/Free Press, 1963); Daniels, pp. 141-53, 172-6; Gilbert, pp. 70-71; Anthony Summers, *Goddess: The Secret Lives of Marilyn Monroe* (London: Victor Gollancz, 1985), p. 59.

prompted them to rush their latest product onto the market in
order to cash in on their investment in rocketry. Both decisions
proved successful. But the field of their success extended well
beyond the studio, for the very existence of these films was made
possible by the popularity of Robert Heinlein's *Rocketship Galileo*
(1947), the text which provided the basic formula for *Destination
Moon* and thereby familiarized a much wider public with the
concept of space travel. The success of both Heinlein's novel and
its screen translation depended in turn on expedient dramatiz-
ations of their social and historical contexts.[69]

Written in the shadow of World War II, *Rocketship Galileo*
rehearsed the successful thwarting of a Nazi plot to establish a
military base on the moon in the form of a myth of youthful
American innovation warding off established imperial design.
Three years later, in the context of a deepening Cold War,
Heinlein's script for *Destination Moon* replaced the Nazi menace
with one from an 'unfriendly foreign power,' and converted the
three adventurous teenagers of the novel into a 'dominant group'
consisting of an inventor, a general, and an industrialist (in
Heinlein's words, 'the just past young, energetic, far-sighted and
dynamic men who are the backbone of American industry'):
Vannevar Bush's guardians of national security. As H. Bruce
Franklin puts it in his study of Heinlein, their flight to the moon
represented 'the triumph of the military–industrial complex.'
That triumph was real enough: *Destination Moon* was released
only a month after President Truman had requested an $11.6-
billion increase in the military budget in order to implement the
terms of NSC–68 within the context of the Korean war. Yet in its
celluloid form it was portrayed, not as the product of state-
financed monopoly capitalism's exploitation of the spoils of war,
but as the achievement of 'an all-star cast of heroes who [were]
the only possible saviors of American society.'[70]

Such a form was, of course, what Hollywood and its audience
required: even a film like Fiedler Cook's *Patterns* (1956) assumed
the business system to be routinely buoyant and marketed its
drama of boardroom warfare as an air-conditioned version of

---

69. Warren, pp. 2-4, 65; Brosnan, pp. 66-7, 74. H. Bruce Franklin, *Robert
Heinlein: America as Science Fiction* (New York: Oxford University Press, 1980), pp.
14, 67-8; Ash, p. 293.

70. Franklin, pp. 76, 97; Robert J. Donovan, *Tumultuous Years: The Presidency of
Harry S. Truman, 1949-1953* (New York: W.W. Norton, 1982), p. 243.

*King Kong.* Equally, as Heinlein's script demonstrates, the means of production of the saviors' rocket were not completely ignored: the race to the moon was in fact specifically presented as 'the greatest challenge ever hurled at American industry.' Nevertheless, the film's presentation of American energy as explicit heroism enabling tacit organization was also what Heinlein himself required in order to resolve – or at least contain – the deepening contradiction of incipient totalitarian society: that unacknowledged tension between the American ethic of self-reliant individualism – which he advocated – and the American experience of disabling anonymity, restricted opportunity, and precarious security, which his advocacy claimed to resist. For if Heinlein's earlier stories dramatized his own nostalgic dream of a nineteenth-century laissez-faire order in the form of lone, superior individuals fighting the erosive encroachments of mass society, monopoly capitalism, and bureaucracy, his post-war script for *Destination Moon* retrieved and updated that hallucination by placing the resources of modern America at the disposal of an heroic self-enforcing crew deployed against the potential encroachments of an 'unfriendly foreign power' – the criminal of Spillane's fictions raised to the planetary scale. In that transposition the hero becomes not only the permitted agent of organized society but also the necessary justification for its experienced constraints: the sanctioned star within an uncertain community. It was an evasive action, but its plot of quick technological victories offered vicarious and thrilling compensation for Americans long impregnated with the heroic vitalist faith which Heinlein's continuing space epic expressed.[71]

Characteristically, the action could be repeated. Between *Rocketship Galileo* and *Starship Troopers* (1959), each of Heinlein's works was characterized by what Franklin calls a 'fever to escape the urbanized, complex, supposedly routinized and imprisoning experience of earth' and by a 'missionary zeal for a colossal human endeavor.' These fictions all went into mass market reprints, spawned other films (*Project Moonbase* in 1953), television

---

71. Sayre, pp. 137-40; Leslie Halliwell, *Halliwell's Film Guide*, 3rd edn (St. Albans: Granada, 1981), p. 763; Irving Pichel, dir., George Pal, prod., *Destination Moon*, script by Robert A. Heinlein, Eagle-Lion, 1950; Franklin, pp. 18-19, 22, 34-49, 99-101; Eric Mottram, 'The Persuasive Lips: Men and Guns in America, the West,' *Journal of American Studies* 10, no. 1 (1976), p. 57; Eric Bentley, *The Cult of the Superman* (London: Robert Hale, 1947).

series (*Tom Corbett: Space Cadet* in 1950) and comic strips, and appeared in papers like the *Saturday Evening Post*. The would-be evacuees 'struggling by science to escape' in *When Worlds Collide* were therefore far from alone. Indeed, their rehearsal of containment by dispersal was promising to become the Rocket State's standard way out. Pynchon's 'virus of Death' had impregnated the culture: now the fever was catching. *Tom Corbett* was only one of a dozen televised science fiction dramas offering space patrols over a decade before *Star Trek*: the *Mittelwerk*'s 'flickering ... space-operetta' was coming to life. When in July 1953 the *Bulletin of the Atomic Scientists* carried the results of a survey which showed that most 10-year-olds felt 'pressurized into social conformity, so that it satisfies them to involve themselves in fantasies of beating the world by vanishing from it,' it only indicated that the intersections of entanglement and evacuation extended even beyond the reaches of the soon to be discovered worlds of juvenile delinquency. The call of Coover's commuters in *The Public Burning* was being repeated across the entire Rocket State (*GR* 296).[72]

To a fortified citizenry strung out between fictions of invulnerability and surrender, the escape clause invented by Heinlein and others – 'the Rocket' itself – offers the most competitive and reassuring of terms: 'an escape from the American nightmare into the fondest American dream – an infinitely expanding frontier.' For where Spillane pushed enforcement and war while Peale preaches adjustment and peace, the rocket promises excitement and pacification. Its launch fuses the urge to conquest with the desire to experience extreme helplessness. It replaces the induced anxiety of the contradictory command structure with the engineered sensation of imminent discharge. And it converts the personalized vantage-points of involvement and inertia, of anarchy and stasis, into one large installation for thrills. In Balint's terms it draws philobats and ocnophiles together. This way out may be no more than the answer to the problems of disposal faced by William Burroughs's protagonist in 'Martin's Folly' and 'The Beginning Is Also The End': an 'escape plan'

---

72. Franklin, pp. 14, 67, 73-4; Brosnan, pp. 291-3; Ash, p. 300; Miller and Nowak, p. 279. In 1951, *Life* estimated that some two million Americans read science fiction of some kind, while the simulated moon rocket installed at Disneyland four years later proved to be among its most popular attractions. See William Sims Bainbridge *et al.*, *The Spaceflight Revolution: A Sociological Study* (New York: John Wiley & Sons, 1976), pp. 198-208; Schickel, pp. 322-5.

designed merely 'to keep this show on the junk road.' It may be just another scene from Gerhardt von Göll's *New Dope*. But it is also the deepest 'shelter in time of disaster' available to the cross-section of evacuees described in *Gravity's Rainbow*.[73] Equally, its entrances may be only dimly illuminated by Oberst Enzian in the depths of the *Mittelwerk* camp and by Tyrone Slothrop in the 'waste regions' beneath the Roseland Ballroom's toilets. But their 'Promise[s] of Space Travel' also lead to the construction and recovery of the 00001, to the network of 'contacts' which spark off the dismantling of Slothrop, the dismissal of Pointsman, the displacement of Gottfried, and the dismemberment of Blicero, and therefore to the redistribution of resources – both hardware and software – which fosters Mossmoon's post-war mass mobilization (*GR* 66-7, 297).

The Rocket's joint offer of absorption and dispersal exemplifies the dynamics of containment within incipient totalitarian society. Its projection of the charismatic complex of hardware and heroics onto the face of the screen fills the Orpheus Theater, activates the elevator, and thus both stimulates and pacifies the invisible roomfuls of fans waiting, between routine boredom and impatient rage, for the show to begin. Its production provides work for crew and cast in their manufacture of novelty, whilst its perception provides leisure for customers in their consumption of spectacle. Its flight-path includes amongst its coordinates a nightmare of destruction and a dream of security, a memory bank of World War and Cold War and an anticipation of victory, a promise of access to the American dream and a precondition of disciplined restraint for survival. Its mission universalizes the program informally canvassed in texts such as Henry Luce's *The American Century* (1941), Anne Morrow Lindbergh's *The Wave of the Future*, and Wendell Willkie's *One World*, formally adopted upon the launching of the 00000 and the absorption of imperialism in 1945, and ostensibly pursued by the American power elite and its agents during subsequent years: the extension of 'Happy-

73. Richard Hofstadter, *The Paranoid Style in American Politics* (1964: New York: Vintage, 1967), pp. 45-6; Michael Balint, *Thrills and Regressions* (London: Hogarth Press/Institute of Psychoanalysis, 1959); William Burroughs, 'Martin's Folly,' *Residu* 2 (1966), quoted in Eric Mottram, *William Burroughs. The Algebra of Need* (London: Marion Boyars, 1977), pp. 72-3, 82; William Burroughs, 'The Beginning Is Also The End,' *Transatlantic Review*, 14 (1963), reprinted in *White Subway* (London: Aloes seolA, n.d. [1973?]), pp. 35-8.

ville' across the remains of 'Pain City,' be they ghetto, mental ward, or 'third world'; the transformation of war into peace, whether by police action, 'pacification,' or 'development'; and the suburbanization of the peripheries, from the soul to the solar system and beyond.[74]

Still, mobilization and mission both require stimulation. In *Destination Moon* it was provided by an 'unfriendly foreign power'; in *When Worlds Collide* by an unfriendly foreign planet. Yet by mid-1957, notwithstanding the austere saber-rattling of Secretary of State John Foster Dulles over Vietnam and Quemoy and Matsu, the combined effects of the death of Stalin and his successors' cautious liberalization, the cease-fire in Korea, the steady recovery in western Europe, and the scrapping of Senator McCarthy had been to restrict further official American investment in the second Red Bank. Between Fiscal Years 1953 and 1956, by shifting resources towards the relatively inexpensive Strategic Air Command, the Eisenhower administration reduced national security expenditures by 20 per cent. At the Geneva summit in July 1955, the President's 'Open Skies' proposal for the monitoring of arms agreements confirmed the desire for a stabilization of relations which he had earlier expressed through the 'Atoms for Peace' initiative of December 1953.[75]

Nevertheless, the sheer extent of the Rocket State's integration ensured a timely reactivation of the account. For in July 1957, the stock market began to slide and the United States entered its worst recession since the war. Wholesale prices declined, as did new orders for durable goods, the combined result of reductions in aerospace defense orders by an administration committed to countering inflation and reducing the budget deficit and of the growing saturation of the consumer durables market. Homes only needed so many telephone extensions, whatever the color.

---

74. Henry R. Luce, *The American Century* (New York: Farrar & Rinehart, 1941); Anne Morrow Lindbergh, *The Wave of the Future* (New York: Harcourt, Brace, 1940; Wendell Willkie, *One World* (New York: Simon & Schuster, 1943), pp. 139-76. See also W.A. Swanberg, *Luce and His Empire* (1972; New York: Dell, 1973), pp. 244-61.

75. John Lewis Gaddis, *Strategies of Containment* (New York: Oxford University Press, 1982), pp. 127-97, 359; Dwight D. Eisenhower, *The White House Years: Mandate for Change, 1953-1956* (London: Heinemann, 1963), pp. 251-5, 443-6, 512-22, 529; Divine, pp. 105-23; Stephen E. Ambrose, *Eisenhower the President. 1952-1969* (London: George Allen & Unwin, 1984), pp. 147-51, 153, 257-9, 264-7, 310-11.

The following month, August 1957, the essential characteristic of what Pynchon called the '*Rocket-cartel*' – that it cut 'across every agency human and paper that ever touched it. Even to Russia' – was confirmed when *Pravda* announced the successful completion of a series of tests on a 6,000-mile-range ICBM. The results were predictable: prospects for investors in Reds now looked rosier – especially for those who had studied earlier banking panics (*GR* 566).[76]

Twenty years before these developments, the shifts in the 'pattern of circumstances' which Hadley Cantril believed to have fostered the *War of the Worlds* panic had included the sharp recession beginning in August 1937 (triggered by the Treasury Department's drive to balance the budget by cutting public expenditure) and a succession of war scares during 1938 (from the *Anschluss* in March through the continuing Czech crisis).[77] The resemblance to the events of 1957 was uncanny. Nor was this all. The previous year Bing Crosby and Frank Sinatra had joked in *High Society* that 'next July we collide with Mars,' only to dive back into their 'swell party' and the struggle for Grace Kelly. When 'next July' brought the recession and next August the Soviet rocket, even Gerhardt von Göll might have been impressed. And if these messages failed to get through to Newport society, the launch of Sputnik I on 2 October followed by Sputnik II on 3 November 1957 (part of the Soviet Union's contribution to the International Geophysical Year) beamed one down for all the world to hear. The reaction was delayed but feverish. Where in 1938 the tensions of imperialism had yielded panic in the streets, now the tensions of incipient totalitarianism yielded anxiety in the media. If, as Alexander Werth reported in *America in Doubt* (1959), Americans were more concerned about a recession than a war, the prospects of their combination proved

---

76. Alexander Werth, *America in Doubt* (London: Robert Hale, 1959), p. 47; Ambrose, *Eisenhower*, p. 460; Eisenhower, *Waging Peace*, pp. 212-13, 216-18; William H. Schauer, *The Politics of Space* (New York: Holmes & Meier, 1976), p. 14; Herbert Stein, *The Fiscal Revolution in America* (Chicago: Chicago University Press, 1969), p. 321; Packard, *Waste Makers*, pp. 11-19.

77. Cantril, pp. 153-60; Charles Jackson, 'The Night the Martians Came,' in Isabel Leighton (ed.), *The Aspirin Age, 1919-1941* (1949; Harmondsworth: Penguin, 1964), pp. 445-7; William E. Leuchtenburg, *Franklin D. Roosevelt and the New Deal* (New York: Harper & Row, 1963), pp. 243-50, 263; James MacGregor Burns, *Roosevelt: The Lion and the Fox* (New York: Harcourt, Brace & World, 1956), pp. 319-28.

intolerable. Impelled further by the crisis at Little Rock in September and by the findings of the leaked Gaither Report in November, the copies of *Life* which reached the Rocket State's elevator in late 1957 and early 1958 quickly helped reactivate the second Red Bank.[78]

The dividends were manifold: demands for mobilization were predictably bullish while interest in the American mission was raised. In all such hectic trading, agile depositors stood to gain. But profits could also be realized in other speculative ventures. Six years earlier the scenario for *When Worlds Collide* had portrayed its global crisis as providing not only a 'magnificent picture opportunity' for the 'take off of a huge rocket,' but also an 'appalling background and suspense for [a] love story.'[79] It, too, proved peculiarly accurate. For in the years following Sputniks I and II a band of leading men would encompass the globe in a battle for hearts and minds. As Julian Smith later recalled in *Looking Away: Hollywood and Vietnam*:

> not since Pearl Harbor had there been such an emotionally power-ful national challenge as that supplied by the launching of Sputnik in 1957. It was a challenge that could reinforce strange alliances.[80]

78. Werth, p. 81; Swanberg, pp. 549-52; *Life*, 18 November 1957.
79. Schauer, pp. 1-2; Warren, pp. 58-9.
80. Smith, p. 202.

# 3

# The Yellow Brick Road

## Invasion USA

> The mass of Americans are not felt as a political reality ... It is only
> when their heart-land, their no man's land, their valley is invaded,
> that one discovers the reality.
>
> — Norman Mailer, *The Presidential Papers* (1963)

> Out! the people want out! – but where is out? The emptiness at
> the edge has inundated the heart, the center is gone, the power cut,
> there's no way in *or* out! ... the communicants, following in the
> footsteps laid down by their heritage and so seized as ever by the
> American go-go-go mania, lurch violently in all directions at once
> ... a movement at once fervid and infinitely varied, yet at the same
> time in a random way rhythmic and predictable.
>
> — Robert Coover, *The Public Burning* (1977)

The launch of the Sputniks in the autumn of 1957 brought forth
an identical reaction in a variety of tongues. In London Conservative Prime Minister Harold Macmillan described the event as 'a
real turning-point in history.' In Peking Chinese Premier Zhou
Enlai considered it 'a new turning-point in the world situation.'
In Paris the eminent physicist Frédéric Joliot–Curie proclaimed
it to be 'a turning-point in the history of civilization.' Further
afield, other voices specified some of the dimensions and potential consequences of this change in direction. In the third world
Sputnik I was greeted by Indian Prime Minister Nehru as a
'great scientific advancement,' while Cairo Radio announced that
the capabilities of its booster undermined the significance of 'all
kinds of pacts and military bases' and would therefore 'make
countries think twice before tying themselves to the imperialist
policy led by the United States.'[1]
Partly as a result of such reactions the prevailing mood in

Washington and across the country was one of alarm and apprehension. If Sputnik constituted a turning-point in history, history appeared to be turning against the United States. Having dismissed as empty boasts a succession of Moscow communiqués announcing forthcoming satellite launches, Americans from Newport to New Mexico awoke one Saturday morning to find the cherished axiom of national technological supremacy in question. As John Gunther remarked, 'for a generation it had been part of the American folklore that Russians were hardly capable of operating a tractor.' Now, in the words of Senator Lyndon Johnson, they had 'beaten us at our own game' and presented the United States with a novel proposition: if the Russians could place a satellite in orbit they might be able to deliver atomic warheads anywhere on earth. Khrushchev had said as much in August after the successful ICBM test. When in the wake of Sputnik I he proclaimed that 'fighter and bomber planes can now be put into museums,' while hinting at 'more things up our sleeve,' the security of Fortress America appeared to have been badly breached. Businessmen who voiced concerns that 'Sputnikitis' might retard consumer expenditure and thereby deepen the recession indicated the nature of the breakdown. Where *High Society* had swelled audiences in 1956, now they saw the shock waves from Sputnik starting to spread across the Orpheus Theater, disturbing its elevator network, closing off its 'invisible rooms,' and threatening to fracture its articulating joints. All was not swell at all: the 'imagination of disaster' rehearsed in *When Worlds Collide* was less of a fantasy than its producers may have thought.[2]

The perception of Sputnik as a 'turning-point,' shared by

1. James R. Killian, Jr., *Sputnik, Scientists, and Eisenhower* (1977; Cambridge, Mass.: MIT Press, 1982), p. 10; Norman Moss, *Men Who Played God: The Story of the Hydrogen Bomb* (1968; Harmondsworth: Penguin, 1970), p. 210; Evgeny Riabchikov, *Russians in Space*, trans. Guy Daniels, ed. Nikolai P. Kamanin (London: Weidenfeld & Nicolson, 1972), p. 147; *New York Times*, 6 October 1957, p. 44, 8 October 1957, p. 13.

2. John Gunther, *Inside Russia Today*, rev. edn (New York: Pyramid Books, 1962), pp. 306, 309; Rowland Evans and Robert Novak, *Lyndon B. Johnson: The Exercise of Power* (London: George Allen & Unwin, 1967), p. 190; Dwight D. Eisenhower, *The White House Years: Waging Peace, 1957-1961* (London: Heinemann, 1966), p. 205; Moss, p. 210; Constance McLaughlin Green and Milton Lomask, *Vanguard: A History* (Washington: Smithsonian Institution Press, 1971), p. 188; William H. Schauer, *The Politics of Space* (New York: Holmes & Meier, 1976), pp. 2, 61-3; *Time*, 2 December 1957, p. 13.

many outside the United States, was translated inside the higher
circles of American society into a more substantive but equally
recurrent metaphor. To Senator Lyndon Johnson on his Texas
ranch, to nuclear physicist Edward Teller on *See it Now* and, in
the opinion of British ambassador Sir Harold Caccia, to much of
Washington society, Sputnik amounted to a second Pearl
Harbor. The peril was no longer yellow but, as the *New York
Times* editorialized after Sputnik II had gone into orbit, the
country once again faced 'a challenge to our nation's existence,'
this time from 'barbarism armed with Sputniks.'[3] Project
RAND's 1946 prediction – that 'the achievement of a satellite
craft ... would inflame the imagination of mankind' – was soon
confirmed. In the US Congress Democratic Senator Mike
Mansfield proclaimed that after years of American complacency
'the chickens [were] coming home to roost,' and called for the
establishment of a 'new Manhattan Project' to wrest missile
superiority from the Russians. Concurrently Republican Styles
Bridges demanded a full-scale Senate enquiry into US defense
capabilities and urged Americans to be 'less concerned with the
depth of the pile on the new broadloom rug or the length of the
tailfin on the new car, and ... more prepared to shed blood,
sweat and tears if this country and the Free World are to survive.'
Much the same message echoed up the elevator shafts and out
across the Orpheus Theater. Clare Booth Luce reported that 'the
beep of the Soviet sputniks is an intercontinental outer-space
raspberry to a decade of American pretensions that the American
way of life was a gilt-edged guarantee of our material superiority.'
Meanwhile husband Henry Luce produced a special 'Sputnik
issue' of *Life* in which his editorial anticipating imminent warfare
preceded a lead article advocating panic, assailing complacency,
and demanding 'a national renunciation of trivialities and a
solemn dedication to serious purpose.' Americans browsing
through this edition between stops clearly faced a most difficult
choice. In the opinion of *Life* they could either watch television,
drive a Cadillac, enjoy lower taxes – or decide to 'live in
freedom.'[4]

---

3. Lyndon Baines Johnson, *The Vantage Point* (New York: Popular Library,
1971), p. 272; Alexander Kendrick, *Prime Time: The Life of Edward R. Murrow*
(London: J.M. Dent & Sons, 1970), p. 404; Harold Macmillan, *Riding the Storm:
1956–1959* (London: Macmillan, 1971), p. 320; Hugo Young, Bryan Silcock, and
Peter Dunne, *Journey to Tranquillity* (London: Jonathan Cape, 1969), p. 53; *New
York Times*, 4 November 1957, p. 28.

It was a choice few Americans recognized and many found hard to understand. For years those selfsame higher circles had been assuring the nation that its patriotic struggle for cars and television was the very embodiment of freedom; now millions who had inched their way to liberty by installments found themselves accused of complicity in its subversion. Not surprisingly, therefore, further down the class structure Sputnik provoked both apathetic dismissals and agonized heart searching, the characteristic reactions of Ellul's propagandized society. Thus *Newsweek* reported 'massive indifference' to the Soviet coup among Americans living in Boston, whereas in Columbus, Ohio, Alexander Werth felt that 'the whole bottom had dropped out of their world' at the news. Likewise, Aneurin Bevan witnessed a frightening obsession with communism during his tour of post-Sputnik America, yet in the mid-western world of the *Milwaukee Journal* it was the performance of the Braves in the World Series which proved worthy of attention, not the behavior of an alleged Russian satellite. Opinion polls provided some corroboration of these antithetical samples. But they also marked out a central complex of ideas to which a clear majority of voters could assent. In January 1958 82 per cent of respondents to an ORC (Opinion Research Corporation) poll agreed that the United States was behind the Soviet Union in the development of advanced weaponry, and 67 per cent felt that Americans had 'been too smug and complacent about our national strength.' At the same time 77 per cent of those questioned told SRC (Survey Research Center) that Sputnik should make 'a difference in what we are doing about the defense of this country,' whilst 61 per cent told a Gallup poll that they endorsed the need 'to pull in our belts and sacrifice for stronger defense.'[5]

---

4. *New York Times*, 8 October 1957, p. 11; Schauer, p. 1; Enid Curtis Bok Schoettle, 'The Establishment of NASA,' in Sanford A. Lakoff (ed.), *Knowledge and Power: Essays on Science and Government* (New York: MacMillan/Free Press, 1966), pp. 183-4; Gunther, p. 309; W.A. Swanberg, *Luce and His Empire* (1972; New York: Dell, 1973), p. 549; Alexander Werth, *America in Doubt* (London: Robert Hale, 1959), p. 105; *Life*, 18 November 1957, pp. 125-8. See also the editorial in *Fortune*, November 1957, pp. 125-6.

5. Green and Lomask, p. 187; Werth, p. 62; Michael Foot, *Aneurin Bevan: A Biography. Vol. Two, 1945-1960* (London: Davis-Poynter, 1973), p. 591; Donald Michael, 'The Beginnings of the Space Age and American Public Opinion,' *Public Opinion Quarterly* 24 (1960), pp. 575-9; Gabriel Almond, 'Public Opinion and the Development of Space Technology,' *Public Opinion Quarterly* 24 (1960), 568; Schauer, pp. 94, 118.

After some consideration the decision appeared to have been made in *Life*'s favor. However, the ORC poll also revealed that while most respondents supported the need for greater effort and more missiles, 71 per cent of them felt that these improvements did not require increases in personal taxation. Guns did not preclude butter. The necessary sacrifices could be made elsewhere. One analyst concluded from these results that reactions to Sputnik had proved to be 'sometimes inconsistent, occasionally rich in non sequiturs, and frequently illogical.' But this was hardly surprising. The tacit assumption that opinion in late capitalist societies should be logical and consistent was a delusion Henry Luce never labored under. The 'Sputnik issue' of *Life* which demanded freedom before freeways also included a 3-page, 4-color advertisement for the 1958 Cadillac, 19 more pages of automobile copy, and a 'kaleidoscopic offering' of other essential defense items like Revlon's Red Caviar Lipstick, Miss Clairol hair coloring and Gordon's Gin. Luce himself sweated blood around the Arizona Biltmore Hotel's private golf course, while his wife tried to pull her belt in at the gilt-edged Elizabeth Arden Spa in Phoenix. In this as in many other cases, inconsistencies in public opinion articulated contradictions in social structure. Where both the publisher and readers of *Life* depended on Cadillacs for their well-being, sacrifice beyond the freeways had to remain a good *idea*; to the extent that it did so, inconsistent attitudes could be effortlessly contained.[6]

More than this, they could be easily reproduced. Those businessmen who at first feared the consequences of the Russian satellite for domestic consumption had little to worry about. Expenditure was not constrained but redirected. Within days of the first launch a Pennsylvania distiller had patented 'Sputnik vodka,' while bars across the country – so the joke went – began advertising 'Sputnik cocktails' (one-third vodka, two-thirds sour grapes). By mid-November an Atlanta restaurant was serving up its first 'Sputnikburger' and a Philadelphia grocer had sold off his stock of undersized potatoes as 'spudniks.' Whilst the *New York Times* proclaimed the 'race for survival,' the *Wall Street Journal* reported numerous toy manufacturers launching crash programs to get new ranges of Sputnik-related novelties into the stores before Christmas. Over on the west coast the filming in Holly-

---

6. Michael, pp. 579-81; Werth, p. 105; Swanberg, pp. 549-52.

wood of Jules Verne's *From the Earth to the Moon* was speeded up
and a number of 1950s science fiction films were re-released –
*When Worlds Collide* amongst them. On the east coast publishers
and bookstores reported increased sales of science fiction titles
and technical manuals. On the Gulf coast the number of UFO
sightings suddenly jumped. Across the entire country sales of
binoculars and telescopes followed suit. Having been exposed at
one of its strong points, those sheltering inside Fortress America
were quickly redeploying themselves to others: evasive action
remained an American speciality.[7]

It was a tactic which American Congressmen in particular
knew all about. Tremors inside the Orpheus Theater registered
faster on the political scales than anywhere else and brought
politicians running with their latest prescriptions for relief,
recovery, and reform. As Stephen Ambrose notes, after Little
Rock and Sputnik had badly damaged two of the central pillars
of the national consensus (faith in the Republicans as the party
most likely to keep the United States out of war and the watch-
word 'Trust Ike'), 'the Democrats were after Eisenhower with a
vigor and enthusiasm previously unknown.' Within hours of
Sputnik I going into orbit Senators Symington, Humphrey,
Kefauver and others had blamed the Soviet coup on the President's lack of leadership and his commitment to fiscal conservatism. They advocated the acceleration and coordination of all US
missile programs in order to repair what they considered 'a
devastating blow to the prestige of the United States as a leader in
the scientific and technological world.' A week later the Democratic Advisory Council of the Democratic National Committee
issued a statement repeating these charges and demands, while
leading Democrats from Adlai Stevenson to Averell Harriman
lent their weight to the chorus of dissent falling on Eisenhower.[8]

The immediate objective was, as former President Truman
put it in mid-October, to 'rub Ike's halo out altogether' in time
for the congressional elections due in November 1958. In the
longer term, however, the intention was to return a Democrat to
the White House in January 1961. To this end the Sputnik issue

---

7. Patrick D. Hazard, 'A Fast Buck on Sputnik,' *The Nation*, 23 November
1957, pp. 379-81; Werth, pp. 63-4, 105-8; Green and Lomask, pp. 188, 206.
    8. Stephen E. Ambrose, *Eisenhower the President. 1952-1969* (London: George
Allen & Unwin, 1984), p. 425; Almond, pp. 568-9; Schoettle, pp. 173-4, 184-5; *New
York Times*, 6 October 1957, p. 42.

was taken up by Eisenhower's critics as the cutting edge of a broader assault on the administration's entire record. Bolstered by the results of opinion polls revealing that the Russian satellites had badly shaken European faith in American technological superiority, increased sympathy for neutralism, and undermined support for NATO and other expressions of US foreign policy, Democratic hopefuls began demanding that the United States respond to the challenges thrown down by Khrushchev by competing with the Soviet Union in a.new type of Cold War for global leadership.[9]

The terms of that war were those of the Rocket State itself. Where Soviet GNP had grown between 1950 and 1958 at an annual average rate of 7.1 per cent, allowing Khrushchev to propose early in 1959 impressive new economic targets and to claim that peaceful economic competition between socialism and capitalism would lead to a Soviet victory 'in a historically short time,' Eisenhower had presided over an economy plagued by recurrent recession, rising unemployment, and growth rates decreasing towards a mere 3 per cent in the late 1950s. Where Soviet propaganda could claim that the USSR had constructed a non-discriminatory society at home and was fostering a post-colonial order overseas with its growing foreign aid schemes, Eisenhower had presided over a nation embarrassed domestically by Little Rock and repudiated abroad by the angry reception Vice-President Nixon received in Caracas in May 1958. Where *Pravda* in November 1957 could describe the Sputniks as 'a vivid expression of the great advantages of the socialist over the capitalist system,' Eisenhower had presided over a country with no satellite coups to cheer and a Defense Department ridden by interservice rivalry over roles and missions in space.[10] The United States therefore had to demonstrate the superiority of the free enterprise system by restoring a high level of economic growth; it had to prove its capacity for justice by making an effective commitment to civil rights; and it had to regain its position as the

---

9. Schoettle, p. 185; Almond, pp. 556-65.
10. See Walter LaFeber, *America, Russia, and the Cold War, 1945-1975*, 3rd edn (New York: John Wiley & Sons, 1976), pp. 199-200; Frederick C. Barghoorn, *Soviet Foreign Propaganda* (Princeton: Princeton University Press, 1964), pp. 166-200; Roy Medvedev, *Khrushchev*, trans. Brian Pearce (Oxford: Basil Blackwell, 1982), p. 158; Wolfgang Leonhard, *The Kremlin Since Stalin* (London: Oxford University Press, 1962), p. 316; Gunther, p. 411.

world's leading scientific power by overtaking the Russians in space. In his memoirs Lyndon Johnson recalled that the United States 'broke out of far more than the atmosphere' with the space program finally spurred by the Russian successes: it also 'escaped from the bonds of inaction and inattention that had gripped the 1950s.' In 1957 Johnson's aims were more precise. Working through his hurriedly constituted Senate Armed Services Preparedness Sub-Committee to produce an 'Inquiry into Satellite and Missile Programs,' his primary aim was to free the electorate from the bonds of President Eisenhower. Sputnik was to be the Republican Party's China: a key by which the doors of the White House might be opened to the Democrats in general and to Lyndon Johnson (already dubbed by his critics the 'space cadet') in particular.[11]

The Democrats' campaign capitalized on Eisenhower's apparent failure to respond to the Soviet challenge. The President remained confident of American security in spite of Sputnik, was unconvinced of the value of a race into space, and therefore refused to allocate vast resources to a program justified in terms of prestige and propaganda, which were factors he considered of secondary importance. Committed to the objective of a balanced budget and anxious about the inflation increased government spending on space might foster; already concerned about the growth of a military–industrial complex which the extension of the Cold War into space could only stimulate; and as a military expert rightly sceptical of arguments equating Soviet satellite capabilities with Soviet military superiority, Eisenhower refused to take up the Democratic gauntlet. Classified information presented in the TCP Report of 1955 and by secret U–2 missions since May 1956 satisfied him that the United States retained overall superiority and had perfectly adequate defense systems to counter any potential threat. As he remarked at his first post-Sputnik press conference, therefore, Soviet space achievements worried him 'not one iota.' There was a much greater danger to the country were he to yield to feverish and ill-informed demands for instant action: the creation of a 'garrison state' whose

11. Johnson, p. 285; Schoettle, pp. 185-6, 226-8; Charles C. Alexander, *Holding the Line: The Eisenhower Era, 1952-1961* (Bloomington, Indiana: Indiana University Press, 1975), pp. 217-18; John Logsdon, *The Decision to Go to the Moon: Project Apollo and the National Interest* (Cambridge, Mass: MIT Press, 1970), pp. 21-2; Evans and Novak, pp. 189-94; Killian, p. 9; Walter A. McDougall, *The Heavens and the Earth* (New York: Basic Books, 1985), pp. 148-9, 151-6.

establishment would undermine the very society and values it was intended to protect.[12]

Eisenhower's assessment of the significance of Sputnik was reiterated by his political and military staff, although in less measured terms. Secretary of Defense Charles Wilson agreed that Sputnik I was a 'nice technical trick' but reassured the American people that 'nobody is going to drop anything down on you from a satellite while you are sleeping, so don't start to worry about it.' Chief of Naval Research Admiral Rawson Bennett briskly dismissed Sputnik as 'a hunk of iron almost anybody could launch.' Special Assistant Sherman Adams announced that the forthcoming American satellite program would be intended to 'serve science,' not to score points in an 'outer space basketball game.'[13] The President's subsequent initiatives – including a nationwide television and radio address reaffirming his confidence in American missile capabilities, a modest increase in Defense Department funds for an accelerated and reorganized missile program, the appointment of MIT President James Killian as Special Assistant to the President for Science and Technology, and the preparation of a National Defense Education Act proposal – were carried out primarily to calm public anxiety and to limit the damage inflicted on Eisenhower himself by his critics in Congress and the media.[14] Together with the much heralded imminent launch of a small, three-and-a-half-pound satellite on board a US Navy Vanguard rocket, part of the nation's contribution to the

---

12. *New York Times*, 17 July 1969, p. 32; Alexander, p. 221; Eisenhower, *Waging Peace*, pp. 210-11, 216-25; Ambrose, pp. 429-30, 433-5; Schoettle, pp. 174-5; Killian, pp. 11-12; Philip J. Klass, *Secret Sentries in Space* (New York: Random House, 1971), pp. 28-9, 38. As McDougall has shown, Democratic critics of the President spoke with only limited knowledge. In fact, the Eisenhower administration had been anything but complacent since the presentation of the TCP Report in February 1955, accelerating research into and development of both missile and satellite technology. However, none of this work could be publicized for security reasons. Such concern with space strategy over propaganda left Eisenhower open to the Democrats' attacks but, as McDougall emphasizes, loss of public face was less important than the loss of potential secret satellite intelligence on Soviet missiles. In practice, the Sputniks proved strategically beneficial to the US since they precluded potential Russian challenges to the legality of American satellite overflight. Since US satellites were more sophisticated in any case, being second to the Russians in no sense meant being second best. See McDougall, pp. 111, 117-24, 128, 221, 224; Paul B. Stares, *Space Weapons and US Strategy: Origins and Development* (London: Croom Helm, 1985), pp. 30-33, 39-40, 51.

13. Killian, p. 10; Johnson, p. 273; *New York Times*, 5 October 1957, p. 2, 16 October 1957, p. 24.

International Geophysical Year, Eisenhower hoped that these actions would close the Sputnik chapter for good. His hopes were to be promptly dashed. On 6 December 1957 the American rocket successfully completed its countdown, rose some four feet off the launch pad, then sank back down and exploded in front of 'swarms of newsmen' and an expectant nationwide television audience. Vanguard, already reduced to the status of rearguard by the Soviet satellites, was lampooned by the press and redubbed 'Flopnik,' 'Kaputnik,' or 'Stayputnik' in what Tom Wolfe later described as 'a hideous cackle of national self-loathing.' The United States, Eisenhower and his advisers notwithstanding, seemed incapable of launching even a small 'hunk of iron.' With no points on the scoreboard, the panic enjoined by *Life* a few weeks beforehand seemed justified after all.[15]

## John F. Kennedy: Filling the Holes in No Man's Land

At 40, he is trim (6 ft., 160 lbs.) and boyishly handsome, with a trademark in [a] shock of unruly brown hair ... He belongs to a legendary family that surpasses its legend: the Kennedys of Massachusetts. He is an authentic war hero and a Pulitzer prizewinning author (for his bestselling *Profiles in Courage*). He is an athlete (during World War II his swimming skill saved his life and those of his PT-boat mates); yet ... Kennedy is [also] recognized as the Senate library's best customer ... No stem-winding orator, Kennedy instead imparts a remarkable quality of shy, sensemaking sincerity. He is certainly the only member of the U.S. Congress who could – as he did – make a speech with his shirttail hanging

14. Eisenhower, *Waging Peace*, pp. 223-5, 244-53; Schoettle, pp. 177-8, 190; Ambrose, pp. 430-31; Killian, pp. 12-30; Logsdon, p. 17; Alexander, pp. 216, 219-21; Green and Lomask, p. 198; Loyd S. Swenson, Jr., James M. Grimwood, and Charles C. Alexander, *This New Ocean: A History of Project Mercury*, NASA SP-4201 (Washington, D.C.: National Aeronautics and Space Administration, 1966), p. 29; Johnson, p. 275; Douglas Kinnard, *President Eisenhower and Strategy Management: A Study in Defense Politics* (Lexington: University of Kentucky Press, 1977), pp. 85-6, 89-93; Stares, pp. 38-42.

15. Swenson, *et al.*, p. 29; Green and Lomask, pp. 206-9; Killian, p. 119; Werth, pp. 158-9; Tom Wolfe, *The Right Stuff* (London: Jonathan Cape, 1979), p. 74. Less than a fortnight later, on 17 December 1957, the US successfully launched an Atlas ICBM; on 31 January 1958 an American earth satellite, Explorer 1, was sent without problems into orbit. See Louis Halle, *The Cold War as History* (London: Chatto & Windus, 1971), p. 346.

out and get gallery ahs instead of aws. Such virtues have made Jack
Kennedy the Democratic whiz of 1957.

— *Time* (2 December 1957)

The impact and consequences of these events were rehearsed in
an essay written by Norman Mailer during the summer of 1960.
For Mailer, as for Alexander Werth and Aneurin Bevan, the
combined effects of Little Rock and Sputnik had 'stunned the
confidence of America into a new night.' If 'the national Ego was
in shock' during October and November 1957, when the
Vanguard rocket blew up in early December the 'fatherly calm of
[President Eisenhower] began to seem like the uxorious melliflu-
ences of the undertaker.' Such reactions only indicated a larger
problem underneath, however. For the Soviet satellites had
penetrated not merely the previously inviolable skies over
Lyndon Johnson's Texas ranch but also the underlying 'no
man's land' of American society: those precarious points of intersection described by the falling 00000 in *Gravity's Rainbow* and
embodied and reproduced after 1945 by Clive Mossmoon. Shock
waves had unhinged the interlocking catch-points of incipient
totalitarian society, disturbed its sprawling joint command structure, elevator network, and lattice of 'invisible rooms,' and
thereby interrupted the Operation's 'nonstop revue' and blurred
the central face on the screen which held the Orpheus Theater in
harness.[16] Where World War II had temporarily produced a
fusion of 'the history of the battlefields' and 'one's own events,'
combining the 'visible ... history of politics' and the 'underground dream life of the nation,' the experiences of the late 1940s
and 1950s which culminated in Sputnik and Little Rock had
progressively widened 'the fissure in the national psyche ... to the
danger point.' Where the absorption of the Oven State within the
Rocket State had provisionally stabilized the personnel structure
of post-war America, recombining the liberated radicles, shells,
nuclei, and solvents of society released across the Zone within the
figure of Clive Mossmoon and reproducing from this prototypical
mass man that characteristic combination of 'civilized man' and
'underprivileged man' engineered by the absorption of imperialism, the cumulative consequences of the hydrogen bomb and
Hollywood, McCarthyism and the FBI, Korea, the *Reader's Digest*

---

16. Norman Mailer, *The Presidential Papers* (1963; St. Albans: Panther, 1976),
pp. 38, 54.

and television had steadily redivided 'the life of politics and the life of myth' to the point where they 'had diverged too far.' By late 1957, as the reaction to the Sputniks revealed, 'there was nothing to return them to one another, no common danger, no cause, no desire, and, most essentially, no hero.'[17]

In Mailer's view – and in this piece there is little sense of sarcasm or irony, presumably because, as Mailer later acknowledged, he was 'engaged in an act of propaganda' – a hero was what America needed. Blicero and the *Führer* principle had once operated as the Oven State's central joint; the Rocket State now required a modernizing axis capable of reintegrating a society of would-be *Führers*; a figure who could, in Mailer's words, 'reach into the alienated circuits of the underground' and 'capture the secret imagination of the country.' Eisenhower did not fit this bill: he had been 'the anti-Hero, the regulator,' a military father figure who had gained access to the screen of the Orpheus Theater during the war for the imperial succession. For all his continued popularity he belonged, in terms of both age and experience, to the past. A new face was needed, a twentieth-century face. Yet even as the tremors from the Sputniks continued to reverberate around the theater, the Operation was proving equal to the challenge. On 2 December 1957 a new face appeared on the cover of *Time* magazine, an attractive alternative to the eyesore of Vanguard: Senator John F. Kennedy of Massachusetts, the 'Man Out Front' whose unannounced presidential campaign was already starting to accelerate. Kennedy was precisely the 'new kind of political leader' Mailer sought. The Sputniks had invaded the American 'no man's land,' and it was Kennedy's intention to prescribe himself as the agent of relief and recovery. They had exposed the divided character of Americans' 'heart-land,' and it 'was Kennedy's genius to appreciate' the essential political truth thereby revealed. As Mailer put it in a later review, by the late 1950s 'the nation could no longer use a father ... [what] we now required was a leading man.'[18]

---

17. Mailer, *Presidential Papers*, pp. 51-5.

18. Mailer, *Presidential Papers*, pp. 37, 38, 53, 55-6, 74; McDougall, pp. 227-30; Henry Fairlie, *The Kennedy Promise* (New York: Doubleday, 1973), p. 81; Victor Lasky, *JFK: The Man and the Myth* (1963; New York: Dell, 1977), pp. 256-8, 262-5; Peter Collier and David Horowitz, *The Kennedys: An American Drama* (1984; London: Pan, 1985), pp. 262-3, 283-7; Norman Mailer, *Cannibals and Christians* (1966; London: Sphere, 1969), pp. 202, 204; *Time*, 2 December 1957; David Halberstam, *The Powers That Be* (London: Chatto & Windus, 1979), p. 353.

Mailer's essay appeared under the title 'Superman Comes to the Supermarket.' Four years later he suggested that perhaps it should have been called 'Filling the Holes in No Man's Land.' The filling would reoccupy the divisions by taking on the form of the territories exposed. The public 'history of politics' and the private 'dream life of the nation' had moved apart in mutual alienation since World War II; Kennedy would reunite them. Since the latter had been 'agitated, overexcited, [and] super-heated' by the film studios of Hollywood into a 'vertical myth trapped within the skull, of a new kind of heroic life,' Kennedy would 'fill the void' by becoming 'the movie star come to life as President.' As the Senator moved towards the Democratic Party nomination in 1960, and then on into the White House, Mailer sensed that 'America's politics would now be also America's favorite movie, America's first soap opera, America's best-seller.' The American 'heart-land' was to be resuscitated by an American heart-throb.[19]

Such a development was entirely consistent with the Operation's post-war recovery from its previous incarnation. In 1945 the Oven State had reached its apotheosis in the figure of Dominus Blicero, who retreated into an isolated and privatized colonial dream world in an attempt to bring the imperial succession to a halt in a personally frozen frame of time and space. In 1960 the Rocket State matured through the figure of John F. Kennedy by releasing, recasting, and relocating that frozen frame within a structure of protracted and collective illusion. In place of a single shot it proffered the persistence of vision; in place of a terminal frame, a continuous film in which the President's acts became 'a panel of scenes in the greatest movie ever made'; in place of a rigid and dominating *Führer*, a fashionable star with an absorbing repertoire of roles.[20]

The metaphor is illuminating but its domain is circumscribed. For as a result of its superstructural focus, Mailer's 'forcing' or 'bending' of reality overshadows the basic material conditions of Kennedy's election. Accession to the Orpheus Theater's screen was neither instant nor magical: it required backstage passes and a thorough knowledge of the ropes, a commitment to house rules

---

19. Mailer, *Presidential Papers*, pp. 37, 52, 57-8; Mailer, *Cannibals and Christians*, p. 203. On Kennedy as a film star figure, see also Fairlie, p. 207; Collier and Horowitz, pp. 292-3.

20. Mailer, *Cannibals and Christians*, p. 203.

and many contacts in the wings. The 'vertical myth' which Kennedy embodied rested on extensive and expensive foundation work: the business of politics. The Kennedy election machine may have possessed a novel public face, but its internal mechanism merely carried to new extremes that combination of large-scale organization, division of labor, and technique characteristic of all major political parties in any late capitalist society. In almost constant operation following the publication of the candidate's *Profiles in Courage* in January 1956, it was driven in turn by what Alexander Hamilton had once described as 'the vital principle of the body politic': money. Kennedy's path to the White House was marked out, not along a yellow brick road (that would be reserved for the voters) but by the greenbacks accumulated over a forty-year period by the leading man's own founding father.[21]

Notwithstanding his lingering resentment against Boston society's earlier exclusivity, by the late 1950s Ambassador Joseph P. Kennedy had come to exemplify the Operation's post-imperial power elite. For the private fortune which fuelled his protégé's rapid ascent, estimated in November 1957 by *Fortune* at between $200 million and $400 million, had been built up as a result of Ambassador Kennedy's involvement in almost every other part of the Orpheus Theater. Since bluffing his way to the Presidency of the Boston Columbia Trust in 1913, Kennedy had made money helping to integrate it (engineering the creation of RKO Studios by taking over the Keith–Albee–Orpheum theater chain for RCA President David Sarnoff), extending its domain and financing that extension (through his real estate deals and banking services), supplying its stock features (producing low budget formula films for high volume formula audiences), and refreshing its customers (first with bootleg and later with licensed liquor). The proceeds of these and other enterprises, notably stock market manipulation and shipbuilding, had in turn allowed Kennedy to join the company of some of the Rocket State's founder members, Franklin D. Roosevelt and Gerard Swope

---

21. Mailer, *Presidential Papers*, p. 74; Jacques Ellul, *Propaganda: The Formation of Men's Attitudes*, trans. Konrad Keller and Jean Lerner (1968; New York: Vintage, 1973), pp. 216-22; Jacques Ellul, *The Technological Society*, trans. John Wilkinson (New York: Vintage, 1964), p. 374; Lasky, p. 230; Fairlie, pp. 39, 41; Herbert W. Schneider, *A History of American Philosophy*, 2nd edn (New York: Columbia University Press, 1963), p. 81.

amongst them. In 1934, having backed Roosevelt as the Operation's savior and provided financial support to the Democrats at a crucial moment, Kennedy had been appointed Chairman of the new Securities and Exchange Commission from which vantage he helped reorganize the banking system in the interests of the corporate liberals. Finally, after his own political prospects had been irreparably damaged by his isolationist sympathies during World War II, his fortune and contacts were channelled into his last and greatest investment: the election of his son to a seat inside Chicherin's administrative state.[22]

Beginning with the 1946 congressional elections, Joseph Kennedy provided financial assistance and strategic introductions to those political bosses, journalists, and business contacts whose support in the wings would prove essential. These foundations in turn furnished the basis for the creation of a political organization unrivalled in its use of the latest techniques in propaganda, analysis, and forecasting. In 1952 John Kennedy's successful senatorial campaign, backed with over half-a-million dollars of family money, was in the words of later campaign biographers 'the most methodical, the most scientific, the most thoroughly detailed, the most intricate, the most disciplined and smoothly working statewide campaign in Massachusetts history – and possibly anywhere else.'[23] Eight years later another $1.5 million had been spent in the run-up to the 1960 Democratic convention on an extension of the electoral machinery. In the race for the White House, John F. Kennedy's campaign made use of in-depth polling provided by Lou Harris, IBM computers, and other machinery designed to predict the reactions of 480 'voter types' (the progeny of Clive Mossmoon's polymerization) and thereby to define policy, and what Theodore H. White described as 'possibly the most complete index ever made of the

22. *Fortune*, November 1957, p. 177; Collier and Horowitz, pp. 27-150 passim; Richard J. Whalen, *The Founding Father: The Story of Joseph P. Kennedy* (London: Hutchinson, 1965), pp. 32-111 passim, 131-5, 375-80; Kees van der Pijl, *The Making of an Atlantic Ruling Class* (London: Verso, 1984), pp. 100-102. Some $100 million of Kennedy's fortune was accumulated as a result of real estate speculation and aggressive rental policies during the Rocket State's wartime accession. See Collier and Horowitz, pp. 149-50; Whalen, pp. 375-80.

23. Herbert S. Parmet, *JFK. The Presidency of John F. Kennedy* (1983; New York: Penguin, 1984), p. 20; Fairlie, pp. 41-5; Whalen, pp. 433, 456; Collier and Horowitz, pp. 223-31, 297-8; Lasky, pp. 263, 406-7; Ralph G. Martin and Ed Plaut, *Front Runner, Dark Horse* (Garden City, N.Y.: Doubleday, 1960), p. 164.

power structure of any national party.' Notwithstanding the
denials and qualifications of his former employees, the intention
of all Kennedy's campaigns was precisely that of a political
machine: efficiently and relentlessly to process the raw material
of voters into the end product of power.[24]

As for his assessment of Kennedy's image, or more precisely
the form of his propaganda, what Mailer rehearsed as the candidate's 'essential unconventionality' was again no more than an
extreme instance of a post-war political convention first glimpsed
by Tyrone Slothrop at Potsdam in 1945: the conversion of politics
into show business, of power into permitted glamor, of the power
elite's agents into decontextualized stars. If some politicians such
as Lyndon Johnson continued to cite legislative activity per se as
a claim to status, then many others – from Estes Kefauver to
Joseph McCarthy – had long since incorporated the cultivation
of image in their political calculations. Equally, if Kennedy in
1960 was to appear as 'the movie star come to life as President,'
then rehearsals had been underway for years (*GR* 380-82).[25]

Kennedy's first campaign in 1946 made great use of his alleged
wartime heroism, an episode much publicized through John
Hersey's article on the PT-109 incident which had been commissioned for *Life* before finally being published in 1944 by the *New
Yorker* and later reprinted in the *Reader's Digest*. It also laid
emphasis on Kennedy's role as the best-selling author of *Why
England Slept*, published complete with a foreword by Joseph P.
Kennedy's associate Henry Luce in 1940. The same methods
were applied during the 1952 Senate race. As Julian Smith later
noted, Luce's 1940 foreword appeared to have set 'in motion an
infernal machine.' In 1953, a year after 300 Washington corre-
spondents voted him 'the handsomest member of the House,'
Kennedy's courtship with Jacqueline Bouvier found its way onto
the cover of *Life*, while their wedding in September again

---

24. Theodore H. White, *The Making of the President 1960* (1961; London:
Jonathan Cape, 1964), pp. 51-7, 93, 102-9, 137; Theodore Sorensen, *Kennedy*
(London: Hodder & Stoughton, 1965), pp. 106-8, 113-21, 184; Lasky, pp. 404-6,
468, 485-6, 542-4. See also Halberstam, pp. 316-23; Arthur M. Schlesinger, Jr., *Robert Kennedy and His Times* (London: André Deutsch, 1978), p. 193.

25. Mailer, *Presidential Papers*, p. 39; Mailer, *Cannibals and Christians*, p. 203;
White, pp. 43-6, 131-5; Peter Biskind, *Seeing is Believing* (1983; London: Pluto,
1984), p. 168; Lasky, p. 243; David Zane Mairowitz, *The Radical Soap Opera* (1974;
Harmondsworth: Penguin, 1976), pp. 127-30; Thomas C. Reeves, *The Life and
Times of Joe McCarthy* (London: Blond & Briggs, 1982), pp. 320-21, 325-6.

featured prominently in Luce's publications. By the time Kennedy's drive for the Democratic nomination began in late 1956, his marketing as a celebrity had been underway for some time. The dimensions of what was to come were indicated during his nationwide campaign, ostensibly for the Stevenson/Kefauver ticket, in October 1956: his car blocked and mobbed by a crowd of Louisville college students screaming that he was 'better than Elvis,' Kennedy found himself even more popular than the singing sensation of the year.[26]

Kennedy's four year long campaign for nomination and election in 1960 maintained this established strategy. As Richard Whalen noted, 1957 'saw the beginning of a publicity build-up unprecedented in US political history' which paid 'little attention to politics' as conventionally understood and instead promoted the junior and still relatively unimportant Senator from Massachusetts as a national celebrity. In 1958 Kennedy started attracting sell-out crowds whose constituents seemed to observers more interested in seeing the candidate and securing his autograph than in considering his policy statements. Polls taken during his Senate re-election campaign the same year suggested that Kennedy's appeal rested to a considerable degree on the public perception of him as an honest, courageous, knowledgeable, and articulate personality. During 1959 this appeal was extended in range. Television, which over the previous decade had become one of the Rocket State's main articulating joints, now reproduced Kennedy as an engaging star in whom millions of viewers could put their faith. The composite image which one critic had recorded in September 1957 – that of a 'clean-cut, smiling American boy, trustworthy, loyal, brave, clean and reverent, boldly facing up to the challenges of the Atomic Age' – was clearly doing the trick. The conclusions were obvious. The Kennedy magic would pull them in. The memorandum distributed early in 1960 by the candidate's chief pollster to the campaign staff ('At all costs Kennedy must avoid being looked upon as a politician') was to be followed to the letter.[27]

---

26. Collier and Horowitz, pp. 121, 158, 180, 229-31, 238; Julian Smith, *Looking Away: Hollywood and Vietnam* (New York: Charles Scribner's Sons, 1975), pp. 78-9; Swanberg, p. 248; Garry Wills, *The Kennedy Imprisonment* (Boston: Little, Brown, 1982), pp. 77-8; Whalen, pp. 423, 432-3; Lasky, pp. 207, 208, 224, 252. On the importance Kennedy placed on Luce's empire of *Life*, *Time*, and *Fortune*, see Halberstam, pp. 351-61.

According to Theodore Sorensen, at least, Kennedy merely avoided the gimmicks and gestures of conventional politicians like Richard Nixon. But as Henry Fairlie points out, this was simply not the case. What he sought to avoid primarily was recognition as a calculating arbiter of factional interests. In a campaign where he needed to command 'the support of an astonishingly wide range of political opinion,' from one end of Mossmoon's Poisson distribution curve to the other, Kennedy presented himself as constrained neither by ideology, nor by interest, nor by party. All roads led to his door. Necessarily, therefore, the substantive content of his speeches was increasingly determined by the formal structure of his propaganda, the recognition patterns which all Hollywood products rely on. This procedure, as Fairlie shows, reached its apotheosis in the stress on the phrase 'New Frontier,' which became a mechanism into which almost anything – from 'unknown opportunities and paths' to 'unfulfilled hopes and threats,' from New Hampshire to Alaska, from 'energy' to 'opportunity' – could be slotted. The approach was in one sense consistent with Kennedy's own political background: his permanent campaigning since 1956 had made him one of the Senate's most notorious non-voting absentees. But it also encapsulated the nature of bourgeois politics within late capitalist America, conditions investigated in Daniel Bell's misnamed *The End of Ideology* in 1959, and implicitly celebrated in Seymour Martin Lipset's *Political Man* the following year. In 1960 it was more difficult than ever to find substantive policy differences between the two main candidates. The nation faced a choice, as Nikita Khrushchev put it, between 'a pair of boots.'[28]

The specific form the campaign took was simply a dramatic elaboration of the critiques offered by other leading Democrats to great effect in the 1958 congressional elections. Kennedy repeatedly told voters that in the last few years – implicitly under Eisenhower and Nixon – the United States had become too complacent and relaxed, too easily satisfied by private self-

---

27. Whalen, p. 445; Lasky, pp. 264, 278, 408-9; *Time*, 2 December 1957, p. 17; Halberstam, pp. 324-7; Marshall McLuhan, *Understanding Media* (1964; London: Abacus, 1973), pp. 352-4.

28. Sorensen, pp. 179-80; Fairlie, pp. 55-60, 77, 79-84, 181-5; Collier and Horowitz, pp. 286, 309; Lasky, pp. 331, 468-70; Robert A. Divine, *Foreign Policy and U.S. Presidential Elections, 1952-1960* (New York: New Viewpoints, 1974), pp. 230-32, 274, 286-7; Alexander, pp. 276-7.

indulgence, and too little concerned about national strength and American prestige. In his speech accepting the Democratic nomination in July 1960 he denounced 'the safe mediocrity of the past' and the 'seven lean years of drought and famine' which had 'withered the field of ideas' and deprived 'too many Americans [of] their will and sense of historic purpose.' Such complacency and inertia, Kennedy went on, were potentially disastrous at a time when the United States and the entire Free World were confronted by a renewed communist offensive designed further to sap American will, undermine American leadership, and thus advance Soviet interests. Continued stagnation on the part of the United States would simply guarantee that the 'periphery of the Free World' would 'slowly be nibbled away' and that consequently the 'balance of power' would 'gradually shift against us.' Fortress America was in this sense like the planet earth in *When Worlds Collide*: a sitting target, a shelter in danger of imminent collapse. Only the election of a new leader with a mandate to 'get this country moving again' could preserve the United States as it approached what Kennedy later defined as 'the hour of maximum danger.' Where Eisenhower merely offered 'platitudes and beatitudes' in an atmosphere of 'soft sentimentalism,' the country needed a leader capable of 'critical and intelligent vigilance' to think out 'tough-minded plans and operations.' At a time when freedom was 'under attack all over the globe,' 'good intentions and pious principles' were no longer a substitute for 'strong creative leadership.' Americans had to 'move outside the home fortress' in order to 'challenge the enemy.'[29]

Kennedy, his speechwriters, and advisers adduced much evidence to substantiate these Orwellian points and advocated specific programs to redress the balance. According to the Democratic candidate, the launching of Sputniks I and II in late 1957 constituted the opening move in the new Soviet offensive. Their radio transmissions beamed communist propaganda across the globe while their delivery vehicles confirmed and

29. Kinnard, pp. 67, 142-3; Arthur M. Schlesinger, Jr., *A Thousand Days: John F. Kennedy in the White House* (Greenwich, Conn.: Fawcett Publications, 1965), p. 64; White, pp. 256-9; John F. Kennedy, *The Strategy of Peace* (New York: Popular Library, 1961), pp. 5, 32, 65; *New York Times*, 8 December 1957, p. 24; Collier and Horowitz, p. 305; Sorensen, p. 178; Bruce Miroff, *Pragmatic Illusions: The Presidential Politics of John F. Kennedy* (New York: David McKay, 1976), pp. 13-14, 41-3; Fairlie, p. 72.

dramatized the existence of the 'missile gap' which Kennedy and other Democrats had been warning about for over a year. Shielded by their 'superior striking ability' the Soviet Union had advanced, not only via 'Sputnik diplomacy' but also through a variety of tactics ranging from intimidation over Berlin to internal revolution in Cuba, from 'indirect non-overt aggression' across Africa to 'limited brush-fire wars' in Vietnam and Laos.[30] Confronted by this offensive, the Eisenhower administration had remained committed to an outdated and inflexible strategy of 'massive retaliation' dictated in part by its commitment to the equally outdated and inflexible dogmas of fiscal conservatism. The consequences had included an economy stumbling into recession for the third time in seven years, leading both to higher unemployment and to a defeat on the Cold War's productivity front – not to mention a host of shortcomings in US defense capabilities which left the nation ill-equipped to turn the tide.[31]

Drawing implicitly and explicitly on the arguments of several informed sources, such as the Gaither and Rockefeller Reports and the writings of senior Army officials like General Maxwell Taylor, the Kennedy team advocated a program of comprehensive rearmament which would provide the capabilities for a new strategy of 'flexible response' designed to replace the old: more strategic, theater, and tactical nuclear weapons utilizing the latest technology to close the missile gap; an expanded and modernized conventional Army prepared to engage in limited combat across the globe; special forces capable of combatting revolutionary guerrillas in their own environment; and more flexible long-term foreign aid programs designed to counter the influence of Soviet inducements in the third world. To finance these initiatives Kennedy advocated an increased annual economic growth rate of 5 per cent, to be stimulated by increased public expenditure as proposed in the writings of advisers like Walt Rostow, Walter Heller, and John Kenneth Galbraith. In moving outside the

---

30. George Orwell, 'Politics and the English Language,' in *Twentieth Century Literary Criticism*, ed. David Lodge (London: Longmans, 1972), pp. 363-4, 366-7; Collier and Horowitz, p. 288; Ambrose, pp. 312-13; Kinnard, pp. 94, 107; Lasky, pp. 458-62; John F. Kennedy, *Public Papers of the Presidents, 1961* (Washington: US Government Printing Office, 1962), pp. 305-6; Miroff, pp. 37, 45, 52; Sorensen, pp. 228-9. On the 'missile gap' see John Lewis Gaddis, *Strategies of Containment* (New York: Oxford University Press, 1982), pp. 182-8; Desmond Ball, *Politics and Force Levels* (Berkeley: University of California Press, 1980), pp. 5-15; Alexander, pp. 224-6, 291.

31. Miroff, pp. 38, 168-9; McDougall, pp. 218-22.

'home fortress,' Americans would necessarily extend its domain.[32]

The substantiating evidence and the formal paradigm slotted together neatly and effectively: just as Pal's disaster movie had pulled in the profits ten years before, so now Kennedy's propelled him into the White House in January 1961. In practice, however, this congruence signified not so much an informed understanding of international affairs as the efficient distribution of propaganda and the creation of myth.[33] Extrapolating from discrete events like the launch of a Soviet ICBM and the revolution in Cuba, Kennedy proposed a unified paradigm of superpower relations whose coherence depended on the depreciation of contradictory evidence (such as signs of a Sino–Soviet split), the misrepresentation of Soviet initiatives (such as Khrushchev's proposals for a *détente*), the overdramatization of peripheral developments (such as the fighting in Laos), and the reproduction of loosely argued and insufficiently informed studies (such as the Gaither Report).[34] Operating within the terms of what Daniel Yergin calls the Riga axioms – that the Soviet Union was a state in the service of an ideology, that communism necessarily sought world domination, that its agents accepted compromises only as temporary expedients within the overall drive for conquest, and that the United States could therefore only confront the Russians from positions of strength – he adduced only that evidence which served to validate these principles. The Soviet offensive originated in the words of presidential candidates rather than behind the walls of the Kremlin.[35]

32. Gaddis, pp. 182-3, 213-14; Alexander, pp. 227-30; Kinnard, pp. 43-7; Alan Wolfe, *The Rise and Fall of the Soviet Threat* (Washington: Institute for Policy Studies, 1979), pp. 17-19, 56-7; Ball, pp. 15-22; McDougall, pp. 212-18; Miroff, pp. 169-70; Fairlie, pp. 30-33; Parmet, pp. 76-7; Herbert Stein, *The Fiscal Revolution in America* (Chicago: Chicago University Press, 1969), pp. 372-86; Jim F. Heath, *John F. Kennedy and the Business Community* (Chicago: Chicago University Press, 1969), pp. 31-9, 94-104.

33. See Ellul, *Propaganda*, pp. 20-21, 31-2, 85-7, 112-16; Roland Barthes, *Mythologies*, trans. Annette Lavers (1957; St. Albans: Paladin, 1973), pp. 117-21.

34. Miroff, pp. 44-8, 65; Alan Wolfe, pp. 18-20; David Horowitz, *From Yalta to Vietnam* (1965; Harmondsworth: Penguin, 1967), p. 352; Adam B. Ulam. *Expansion and Coexistence: Soviet Foreign Policy 1917-1973*, 2nd edn (New York: Praeger, 1974), pp. 605-6, 621; Medvedev, pp. 145-9.

35. Daniel Yergin, *Shattered Peace: The Origins of the Cold War and the National Security State* (1977; Harmondsworth: Penguin, 1980), pp. 11-13, 35-9, 168-70; Miroff, pp. 44-5. On the subsequent post-election abandonment of the fiction, see Ball, pp. 88-104; *New York Times*, 7 February 1961, p. 1.

Nevertheless, the theory seemed plausible. On the one hand the quality of the evidence was less important than its quantitative effect: as Jacques Ellul notes, in a propagandized environment a 'surfeit of data' can actually preclude rather than foster thought. On the other hand the effectiveness of Kennedy's message relied on a lot more than specific information: as Roland Barthes points out, the 'mode of presence' of a mythical concept is not simply literal but memorial.[36] Regardless of the dynamics of Laotian society or of Russian technical shortcomings, Kennedy's theory of a communist offensive rang true because many voters had heard – or seen – it all before. When Kennedy cast himself in the role of Winston Churchill, turning Eisenhower implicitly into Neville Chamberlain, the Soviet Union automatically took on the mantle of Nazi Germany in an equation which had gained considerable currency as the Cold War deepened through the concept of 'Red Fascism.' Likewise, when Kennedy told voters that 'the man who merely builds a fortress of his home will always find in the end that the enemy has devised a way to get in the back door,' that same enemy took on the mantle of other, more fantastic or criminal evils in an equation which Hollywood had invoked repeatedly over the years.[37] If the election of Kennedy had promised, as Mailer argued, to reunite the public 'history of politics' and the private 'dream life of the nation,' that integration included an interlacing of recognition patterns. The 'progressive *knotting into*' proposed at the outset of *Gravity's Rainbow* involved the further ingestion of *New Dope*, as Gerhardt von Göll had long dreamed. American policy makers were not the only ones to be guided by 'lessons of the past.' The 'imagination of disaster' could be invoked across the entire Orpheus Theater (*GR* 3, 527, 745).[38]

Moreover, the theory and its lessons yielded further dividends. For if Kennedy's election amounted to a formal recognition of the communist offensive, it also implied acceptance of the final plank

---

36. Ellul, *Propaganda*, p. 87; Barthes, p. 121.

37. Miroff, pp. 14, 39; Collier and Horowitz, pp. 121, 240, 288-9; Lasky, pp. 460-61; Parmet, p. 143; Les K. Adler and Thomas G. Paterson, 'Red Fascism: The Merger of Nazi Germany and Soviet Russia in the American Image of Totalitarianism, 1930s-1950s,' *American Historical Review* 75 (1970), 1046-1064; Nora Sayre, *Running Time: Films of the Cold War* (New York: Dial Press, 1982), pp. 25-30, 79-90.

38. Mailer, *Presidential Papers*, p. 51; Ernest R. May, *'Lessons' of the Past: The Use and Misuse of History in American Foreign Policy* (New York: Oxford University Press, 1973).

in his platform: the demand for 'strong creative leadership' to counter that offensive. Since American leadership was specifically under attack, heroic statesmanship on the part of the President was not only sanctioned but required. Not surprisingly, the argument suited its prime exponent. For whereas he was neither interested in nor well-equipped to handle domestic issues, Kennedy, like Robert Heinlein, found in the collision of the Free World and the communist bloc a dramatic setting for great personal acts: a stage, unimpeded by the conventional constraints of domestic partisanship, economic limitation, or institutional fetters, on which to make history.[39]

History, however, was no mere private indulgence (as it became for Blicero in *Gravity's Rainbow*). Greatness also required public recognition: the audience had to be made to feel involved. At the same time, therefore, Kennedy argued that the Soviet challenge also demanded much of the American people. They had to be prepared 'to sacrifice, to give, to spare no effort' if freedom were to prevail. This did not mean, however, that they should take the stage of history themselves: for all the rhetoric of his inaugural address – 'ask not what your country can do for you; ask what you can do for your country' – Kennedy entertained few concrete proposals for active mass involvement. Instead, beyond the workplace at least, theirs would be a sacrifice akin to that demanded by *Life* in late 1957: a fantasy of exertion – *The Power of Positive Thinking* in secular form – which in practice amounted to little beyond applause, votes, and (lower) taxes. The voters under Kennedy would *spectate*. Like improved and modernized Slothrops, they would enjoy a vicarious identification with their glamorous stars, a rapturous genuflection before the drama of Camelot.[40]

Popular activism was disparaged for two reasons. First, notwithstanding his reductive conception of international relations, Kennedy considered subjects such as foreign policy too complex for the general public to understand and too delicate for them to influence. Issues where mass action predominated – such as civil rights – were issues he might not be able to control. Politics was, in any event, the domain of an elite.[41] Secondly, public involvement implied public debate and therefore potential

---

39. Miroff, pp. 11-13, 67; Halberstam, pp. 386-7.
40. Fairlie, p. 83; Miroff, pp. 25-8; Sorensen, p. 248; Halberstam, pp. 387-9.
41. Miroff, pp. 24-7.

divisiveness. A fervent nationalist elected by a slender majority and with only tenuous congressional support, Kennedy was more concerned with the need to maintain his popularity across the political spectrum than with fostering widespread critical sense amongst voters. As Bruce Miroff notes, 'he wanted their approval and their support, but not their judgement òr their action.' The mass of men and women who had strung themselves out along the interface of inertia and involvement during the Eisenhower era would therefore be reintegrated through passive participation: held, in Henry Fairlie's words, in 'a constant state of expectation'; crammed into the stalls of Pynchon's Orpheus Theater, balanced on the edge of their seats.[42]

What Henry Fairlie labelled the politics of expectation thus revealed two dimensions of the Kennedy *Weltanschauung* whose intersection would structure the Rocket State's apotheosis. First, Kennedy subscribed to the tenets of heroic vitalism as expressed in the works of Carlyle, Spengler, Shaw, and others. Where Carlyle's *Heroes and Hero-Worship* (1841) assumed that evolutionary laws governed both natural and civil history, that the entire universe followed 'harsh biological principles,' and that these determined an organic order to society and social change – encapsulated within naturalist metaphors – whose disruption amounted to illness, Kennedy saw the Cold War as a conflict in which only the fittest would survive, proposed the United States as the 'wave of the future,' and considered communism an outdated and debilitating system: in his own words, 'as old as the Pharaohs,' and in Walt Rostow's 'a disease of the transition' between the primitive and civilized worlds.[43] Carlyle and Spengler considered the hero to be a figure who 'cooperated with the real tendency of the world,' who acted as the 'instrument of history and progress' (in Spengler's words, of 'destiny'), and who was therefore 'the highest product of the evolutionary process.' Kennedy's love of the novels of John Buchan and of biographies of great heroes indicated a comparable belief in the supremacy of heroic action. Carlyle, Spengler, and Shaw divided mankind into two groups ('the many fools and the few wise, the shiftless major-

42. White, pp. 350-51; Parmet, pp. 90, 205; Miroff, p. 24; Ellul, *Propaganda*, pp. 25-7; Fairlie, p. 8.

43. Eric Bentley, *The Cult of the Superman* (London: Robert Hale, 1947), pp. 55-7, 103-4, 234; Miroff, p. 52; Kennedy, *Public Papers, 1961*, pp. 3-5; *New York Times*, 21 September 1960, p. 1; W.W. Rostow, *The Stages of Economic Growth*, 2nd edn (New York: Cambridge University Press, 1971), pp. 162-4.

ity and the energetic minority'), argued that the latter were the agents of order and security within the amorphous flux of events, and therefore insisted that great men should rule and that the masses had to be molded to the curve of their destiny in the interests of civilization. Kennedy associated his band of brothers with the chosen few (the 'best and the brightest,' as David Halberstam called them) and enthusiastically took upon himself the great man's burden of bringing destiny to the masses, from Selma to Saigon. For all of them history was the history of great men; for Kennedy it was a history he would reproduce.[44]

Secondly, and in consequence, Kennedy required this molding of the masses to produce an audience of followers who were united. In his book on the sources of World War II, *Why England Slept* (1940), Kennedy had argued that in the late 1930s Britain, as a democracy, had faced severe difficulties in organizing its national resources for war against the totalitarian Nazi state, where national unity was enforced. Its only salvation lay in what he called 'voluntary totalitarianism,' the willful sacrifice of personal and factional interests to the greater public good. The theory proved a convenient and durable alibi for the unity he required of a modern, socially diverse democracy fragmented into millions of privatized stalls. Twenty years later he campaigned and held office in what he considered congruent conditions. Since the United States was effectively at war with the Soviet Union, they too needed to sacrifice partisan interests in order to counter the enforced unity of communist totalitarianism. Factional self-serving, whether by US Steel or civil rights activists, would only foster divisions, provide easy targets for Soviet propaganda, and thereby damage the United States in the eyes of the world. Since the enemy was pushing forward at all points, the New Frontier had to be manned and advanced by every American in all theaters, from the schoolroom to the missile silo. As in the late 1930s, the 'national purpose' could not 'permit group interests to interfere with its fulfillment.' The government and the people had to act as one.[45]

As Fairlie argues, Kennedy's politics of expectation raised a

44. Bentley, pp. 17, 18, 166, 234; Clive Bush, *The Dream of Reason* (London: Edward Arnold, 1977), p. 46; Fairlie, pp. 116-17; David Halberstam, *The Best and the Brightest* (London: Barrie & Jenkins, 1972).

45. Fairlie, pp. 69-70, 109, 200; John F. Kennedy, *Why England Slept* (1940; London: Sidgwick & Jackson, 1962), pp. 147, 160-61; Miroff, pp. 177-80; Schlesinger, *Robert Kennedy*, p. 299.

pair of related questions. On the one hand, 'what part of the private life of an individual, under such a regime, was to be permitted to remain separate from the public domain?' On the other hand, what happened to the notion of representation in a society where the total fusion of private and public spaces was a precondition of survival? The issues are raised but left undeveloped. Fairlie's conservatism facilitates his description of the contradictions within the Kennedy promise but constrains his ability to analyze them inside their historical field. However, once handled in terms of the changing relationships among private, public, and social spaces as contingent forms engineered and transformed by the historical processes of capitalism, what Fairlie considers geological faults appear as not merely the logical consequences of liberalism but the very apotheosis of incipient totalitarian society. As H.T. Wilson argues, inside what Pynchon projects as the Rocket State the private space *does* tend to be reduced to (and reproduced as) the socially determined functions of consumption, spectatorship, and passivity, whilst the public space *does* tend to be reduced to the socially determined functions of apolitical administration and management; in each case, that is, to behavior sanctioned by class rule in the interests of the development of existing relations of production, consumption, and exchange. If in the process private life is reproduced as insatiable identification while the behavior of public representatives carries 'the contract theory just about as far as it can go,' then such contingencies reveal geological faults or contradictions in the material bases of social structure, not merely in the superstructural reaches of the stars.[46]

The terms of the Kennedy promise did not constitute dehistoricized magical formulae to be followed, nor were they unilaterally imposed. Rather, they articulated the tensions of the ordinary men and women who had filled the pages of Coover's *The Public Burning* and Riesman's *The Lonely Crowd* and at the same time proffered a way out. In Pynchon's terms they reached into and drew upon the structural foundations of the Orpheus Theater, reactivating its elevator network, reopening its 'invisible rooms,' and refilling its joints with the descendants of the Rocket State's developmental and prototypical units. Where Slothrop had been

---

46. Fairlie, pp. 5, 83-4, 87, 106-7; H.T. Wilson, *The American Ideology: Science, Technology and Organization as Modes of Rationality in Advanced Industrial Societies* (London: Routledge & Kegan Paul, 1977), pp. 171-96.

seduced by fantasies of security and satisfaction into a subordinate role within the power elite's 'showbiz' routines at Potsdam, his scattered shells were now drawn in by an identical act from one of Harry Truman's most glamorous heirs. Where Pointsman had developed Skinnerite behavioral engineering techniques only to see them appropriated by agents of the Operation as a result of his inefficiency and inability to raise financial support, one of the Operation's wealthiest families now efficiently employed such techniques on the offspring of his former subject. (As I.F. Stone put it in November 1960, 'Kennedy's campaign gives the impression of being run by technicians who repeat key words on the basis of decibel response.') Where Blicero had sacrificed Gottfried in the interests of the Operation's survival, now Kennedy asked the progeny of that discharge to sacrifice themselves to the selfsame end. Where Greta Erdmann had surrendered to the products of Imipolex in Pynchon's satire on the rise of organized commodity fetishism, uncounted consumers now fell under the spell of the newest and most attractively wrapped article available. Finally, where the production of Clive Mossmoon had indicated the emergence of incipient totalitarian society's basic unit of containment and reproduction, a synthesized, self-regulating behavioral figure in whom private and public spaces were fused, now the Kennedy campaign and presidency proposed a society in which masses of Mossmoon's offspring might mold and mobilize themselves. Within months of the leading man's election in late 1960, the template for that molding and mobilization, the 'way out' proposed and pursued across *Gravity's Rainbow* and the Rocket State, would be pressed into service (*GR* 486-7).[47]

## To Capture the Imagination of the World

'Where does it go?'
'Wherever we tell it to.'
'May I fly in it someday? I'd fit inside, wouldn't I?'
She asked impossible questions. 'Someday,' Pökler told her.
'Perhaps someday to the Moon.'
'The *Moon* . . .'

— Thomas Pynchon, *Gravity's Rainbow* (1973)

---

47. Stone quoted in Fairlie, p. 75.

> In years to come history may record that one of the greatest
> triumphs of the space program of the 1960s was the basis it
> provided for reaching accommodations with old adversaries.

> — Lyndon Johnson, *The Vantage Point* (1971)

The theory of voluntary totalitarianism outdistanced the volunteers, both qualitatively and quantitatively. If Churchill could be
hung in effigy by Kentish miners during World War II and
turned out of office within weeks of victory, the prospects for
national discipline, in a period when *Time* was sandwiching its
post-Sputnik cover story on Kennedy between advertisements for
Gordon's Gin and the ubiquitous 1958 Cadillac, were slim. As
Arthur Schlesinger, Jr., noted, the President faced difficulties
rallying potential troops in conditions where neither armed
conflict nor 'national economic collapse [were] making his constituents clamor for action.'[48] In 1960, for all his efforts, more people
voted against him than for him, leaving his popular mandate
tenuous and his political influence, particularly in Congress,
restricted. Kennedy might impersonate Churchill but he could
never reincarnate him: the power structure had displaced
military struggle from center stage and with it the spirit of
Dunkirk and the blitz. Churchill's had been in effect the last
stand of the British imperial wing of the Operation just as Hitler's
had been for the German; what absorbed Pointsman at the
nostalgically named (final) resort of Ick Regis was directly related
to what Blicero surrendered to on Luneburg Heath: the knowledge that the '*l'état c'est moi* frame of mind' had seen its finest
hour (*GR* 272-7, 721-4, 749-60).[49]

In the incipient totalitarian society mapped out in Pynchon's
epic, no single figure could shelter all the polymorphous, pluralis-
tic faces in the crowd. As Henry Fairlie notes, people released
from the demands of hot war 'no longer have a single objective.'
Some want color television, others seek out libraries; some want
the Cadillac and the Gordon's Gin, others need hospitals. Nor, as
Kennedy himself argued, could any static Fortress America
accommodate these diverse objectives: if Americans wanted
freedom they would have to 'move outside the home fortress' to

---

48. Angus Calder, *The People's War: Britain, 1939-1945*, 2nd edn (London:
Panther, 1971), pp. 282, 457, 497-512; Kenneth O. Morgan, *Labour in Power, 1945-
1951* (1984; Oxford: Oxford University Press, 1985), p. 286; *Time*, 2 December
1957, passim; Schlesinger, *A Thousand Days*, pp. 660-61.

49. White, p. 385; Miroff, pp. 14-15; Schlesinger, *A Thousand Days*, pp. 84-5.

find it; if they wanted security they would have to serve on the New Frontier. However, as in 1957, the integrated nature of the Rocket State ensured that the Operation would respond to such demands. For within a few weeks of his inauguration, the dramatic reinforcement of two fronts in the purported Soviet offensive led Kennedy to place a new and more accommodating object at the center of the national stage.[50]

On 12 April 1961 the Soviet Union successfully capitalized on the achievements of the Sputniks by launching Yuri Gagarin into orbit, whereupon Kennedy quickly found himself tarred with the same brush he had previously used on Eisenhower and Nixon. A few days later the gloomy prognoses of his first State of the Union address were confirmed at the Bay of Pigs.[51] Whilst opinion polls showed the electorate rallying behind the President, Kennedy believed that compensatory action was necessary. That action came quickly. On 21 April a new objective was floated in the pages of *Life*, the Orpheus Theater's house magazine: one which Kennedy knew would introduce new figures into his calculations, one which he hoped would involve the electorate in a variety of ways on the New Frontier, and one which he believed would offer something even to those viewers who remained unmoved by his face on their screens. On 25 May 1961, in an address to a joint session of Congress on 'Urgent National Needs,' the President spelled out the nature of that objective. Confronted by a relentless Soviet offensive which presented the United States with 'an extraordinary challenge,' it was time to move outside the fortress: 'time for a great new American enterprise' whose results could 'in many ways ... hold the key to our future on earth.' The enterprise involved an invasion which would succeed where the Cuban exiles had failed. The United States, Kennedy went on, 'should commit itself to achieving the goal, before this decade is out, of landing a man on the moon and returning him safely to earth.'[52]

As *Gravity's Rainbow* suggests, the new objective had been a coming attraction at the Orpheus Theater for years. In the dying days of the Oven State, Franz Pökler had told his daughter that the rockets he worked on might carry them 'someday to the

---

50. Fairlie, pp. 76, 112.

51. Logsdon, pp. 100-112; *New York Times*, 17 April 1961, p. 5, 20 April 1961, p. 9.

52. Schlesinger, *A Thousand Days*, p. 273; Halberstam, *Powers That Be*, p. 385; *Life*, 21 April 1961, pp. 26-7; Logsdon, pp. 106-7, 127-9.

Moon' and thereby to a world free of wars and nation-states: a universal 'Happyville' beyond the Zone of imperialism's dissolution. For her part Ilse Pökler had hoped to 'fit inside' their engineered arcs and to be carried to the shelter of family life in the Sea of Tranquillity: the most untouchable place of all and the destination of Apollo XI in 1969. Thereafter Pökler's historical template, Wernher von Braun, had continued his classified work on the salvaged V-2 hardware in New Mexico and Texas, developing from their foundations the Redstone rocket and, after President Eisenhower's endorsement of the von Neumann and TCP reports, a series of launch vehicles (from Jupiter to the Saturn V) which would convert Pökler's dreams into reality. Simultaneously, *Destination Moon, When Worlds Collide*, and their progeny had been performing more conspicuously as what Jacques Ellul terms 'pre-propaganda' for over a decade: 'reducing prejudices and spreading images apparently without purpose,' filling in the theater's paying public with the Rocket State's software, fuelling the citizens' imaginations in preparation for their assisted assent (*GR* 400, 410).[53]

The new figures Kennedy had in mind were both financial and personal, and involved closely related calculations. First, a program costing perhaps as much as $40 billion required the support of a packed house in more ways than one. Thus Kennedy's 25 May speech only included the proposal for a moon landing after careful lobbying of influential Representatives and Senators in early May had confirmed the existence of an established bipartisan constituency in Congress; and their support was only drawn upon after the administration, in the person of Vice-President and Space Council Chairman Lyndon Johnson, had gained the approval of key military, financial, and industrial interests surrounding the legislature: from the Pentagon to the Bureau of the Budget, from Johnson's old ally and fund raiser George Brown (of the Houston construction firm Brown and Root) to Johnson's old ally and fund raiser Frank Stanton (President of CBS and Chairman of RAND).[54] As a result

---

53. Frederick I. Ordway III and Mitchell R. Sharpe, *The Rocket Team* (London: Heinemann, 1979), pp. 344-89; Ellul, *Propaganda*, pp. 15, 30-31.

54. Logsdon, pp. 24-6, 108-9, 114-15, 120-21, 141; Johnson, pp. 280-81; Halberstam, *Powers That Be*, pp. 438-40, 515; Robert A. Caro, *The Years of Lyndon Johnson: The Path to Power* (London: Collins, 1983), pp. 369 ff; Herbert Schiller, *Mass Communications and American Empire* (New York: Augustus M. Kelley, 1969), pp. 55-6.

of these consultations Kennedy's proposal received its first material backing in only ten weeks. Having passed unopposed through the Senate and by a 6:1 margin in the House, on 7 August Congress 'approved his requests, almost without a murmur' of dissent, adding over $400 million to NASA's fiscal 1962 budget and initiating a phase of rapid expansion which would continue unabated until 1966.[55]

Such overwhelming legislative backing was bolstered by one of the second sets of figures in Kennedy's calculations. For on 5 May 1961 around forty-five million Americans had watched the successful sub-orbital flight of astronaut Alan Shepard inside a Mercury capsule on live television, whilst three days later some 250,000 had turned out to cheer the new hero along Pennsylvania Avenue on his way to address the assembled members of Congress. Three years earlier President Eisenhower had accepted the establishment of the National Aeronautics and Space Administration in order to accommodate the continuing demands of the Democratic majority in the legislature. Having signed the Space Act in late July 1958, he had also reluctantly endorsed the new organization's initial manned space flight proposal – Project Mercury – in mid-August. Now Kennedy could watch the returns on his predecessor's investment piling up. As the official NASA historians of the Mercury program later wrote, Shepard's opening salvo in the US counterattack clearly sparked 'a formidable change in the public's attitude towards the space program.' Millions of Americans appeared as keen to be involved as the politics of expectation demanded. For their part both President and Congress quickly recognized a potential well of support waiting to be tapped.[56]

For Kennedy himself, Shepard's flight constituted the first stage of his personal political comeback following the reverses suffered at the hands of Gagarin in space and Fidel Castro in Cuba. At the same time it marked the beginning of a relationship which, if not quite the 'love story' proposed in the scenario for *When Worlds Collide*, at least resulted in an enduring marriage of convenience. The engagement was first announced to the

55. Logsdon, p. 129; Jay Holmes, *America on the Moon: the Enterprise of the Sixties* (Philadelphia: J.P. Lippincott, 1962), pp. 199-205; Sorensen, p. 526.

56. Swenson, *et al.*, pp. 101-2, 341, 362; Hugh Sidey, *John F. Kennedy. Portrait of a President* (1963; Harmondsworth: Penguin, 1965), pp. 166-7; Holmes, p. 93; Tom Wolfe, pp. 268-9; Logsdon, pp. 23, 30; *New York Times*, 6 May 1961, p. 1.

cameras on the White House lawn on the morning of 8 May when Kennedy welcomed the Mercury astronauts, and thereafter in the Rose Garden where he presented Shepard with NASA's Distinguished Service Medal. By the time the happy couple had parted, with Shepard leaving for the Capitol by way of a motorcade which attracted a bigger crowd than the President drew at his inauguration, Kennedy had confirmed for himself what a secret memorandum from NASA Administrator James E. Webb and Defense Secretary Robert McNamara would tell him later that day: 'It is man, not merely machines, in space that captures the imagination of the world.' If he could not monopolize public attention directly, Shepard's flight indicated that the President might profitably share the limelight with other celebrated figures whose apolitical, non-partisan stardom could both attract attention and deliver votes. Successful astronauts could be cast as successful leading men.[57]

As Pynchon's novel suggests, in order to maintain incipient totalitarian society's joint command structure, the face on the screen of the Orpheus Theater needed to be both variegated and flexible. The space epic ordered by President Kennedy in May 1961 filled the bill precisely. Part war film, part science fiction spectacular, part costume drama, part tragedy, this elaborate remake of *Destination Moon* paraded a modernized all-star cast of heroes who were, like the Imipolex of Pynchon's novel, ready to take on anything. It projected the face of the President alongside the faces of the astronauts. It integrated a band of leading men whose exploits would bring relief from the constraints imposed by an updated and still 'unfriendly foreign power.' It reproduced a premier collection of 'showbiz types' whose joint appeal would absorb audiences both at home and around the world. Like many other box office successes, like *High Society* itself, it offered something for all the family (*GR* 380).

While Kennedy's new program appeared to borrow its basic plot from *Destination Moon*, however, some of the qualities of its leading men bore a closer resemblance to those celebrated in one of Robert Heinlein's more recent texts. For in *Starship Troopers*, published soon after the seven Mercury astronauts had been

57. Swenson, *et al.*, pp. 361-2; Sidey, pp. 166-7; Bill Warren, *Keep Watching the Skies! American Science Fiction Movies of the Fifties. Vol. 1: 1950-1957* (Jefferson, N. Cal. and London: McFarland, 1982), pp. 58-9; Logsdon, pp. 122-5; Young, *et al.*, pp. 114-15.

selected in 1959, the dominant elements in Kennedy's nascent *Weltanschauung* are condensed into a fantasy of infinite conflict in which an elite force of highly trained warriors is deployed to defend the 'Terran Foundation' against a rising tide of revolution directed by 'The Bugs.' These special forces wage the permanent wars required by the demands of earth's galactic destiny, ridding the cosmos of the agents of disorder whose 'total communism' is enforced via the dictatorship of the hive, and spreading the benefits of a civilization in which voluntary totalitarianism is a precondition for survival and therefore compulsory. Within the Federation only the elite who are willing to devote their lives to the state can govern and only those enlisted in the ranks may vote; history is translated as the lessons of warfare whilst moral philosophy articulates the law of the jungle.[58]

The social structure dramatized in *Starship Troopers* is riddled with the same self-evident contradictions that characterized Kennedy's Cold War combat drama. But, as with the White House best-seller, these contradictions are contained within the structure of incipient totalitarian society, not simply in Heinlein's text.[59] They are therefore articulated not only in the language of Kennedy's New Frontier but also in the production, distribution, and consumption of that frontier's pioneering spirits, the astronauts themselves. These leading men will prove, like the President, to lead nothing. And it is to their operation we turn next.

58. Robert Heinlein, *Starship Troopers* (New York: G.P. Putnam's Sons, 1959).

59. H. Bruce Franklin, *Robert Heinlein: America as Science Fiction* (New York: Oxford University Press, 1980), pp. 111-17; Brian Ash (ed.), *The Visual Encyclopedia of Science Fiction* (London: Pan, 1977), p. 108.

**The Rush Hour.**
On 16 December 1944 a V–2 rocket made a direct hit on the Rex cinema in Antwerp.

**Lie Still and Be Quiet.**    Inside the packed theater 296 British soldiers and perhaps as many more civilians were killed. Hundreds more were injured.

**Roles Reassigned.**    Retreating from the underground *Mittelwerk* factory at Nordhausen, SS troops abandoned the V-weapon production lines to advancing Allied forces.

**The Promise of Escape.**    Salvaged, patched up, and repainted, a captured V–2 is prepared for launching during the Allied *Backfire* test series at Cuxhaven in October 1945.

**Way Out West.**   In late May 1945 another V–2 headed towards Antwerp, this time to become the centerpiece of an exhibition of war materials opened by the US Ambassador to Belgium. Thereafter, the rocket was shipped across the Atlantic to the US Army's White Sands Proving Ground in New Mexico.

DANGER
V 2
KEEP CLEAR

**The Happy Couple.** Alan Shepard, the first American in space, is presented with NASA's Distinguished Service Medal by President Kennedy at the White House on 8 May 1961.

John F. Kennedy Library

Popperfoto

**Start the Show.** Thousands of commuters at Grand Central Station, New York, find the time to watch television pictures of the Atlas rocket carrying John Glenn and *Friendship 7* into orbit on 20 February 1962.

**Spheres of Influence.** Watched by President Kennedy and Vice-President Johnson, John Glenn signs the National Geographical Society's 'flyers and explorers globe' in the Oval Office of the White House on 26 February 1962.

**Spheres of Affluence.** Visitors gather round the 140-foot high 'Unisphere', symbol and centerpiece of the 1964-5 New York World's Fair.

Steep ramp of '124 is a major safety problem, but one that has to be lived with. Note protective coverings on thrust chamber, fin tip, radome.

# "TOGETHERNESS"

**Thomas H. Pynchon, Bomarc Aero-Space Dept., Boeing Airplane Co., Seattle**

Airlifting the IM-99A missile, like marriage, demands a certain amount of "togetherness" between Air Force and contractor. Two birds per airlift are onloaded by Boeing people and offloaded by Air Force people; in between is an airborne MATS C-124. One loading operation is a mirror-image of the other, and similar accidents can happen at both places. Let's look at a few of the safety hazards that have to be taken into account when Bomarcs are shipped. . . .

In the July 1960 issue of *Aerospace Safety*, mention was made of the second Air Force-Industry conference on missile safety; and of plans to create Air Force-Industry Accident Review Boards. If future emphasis is to be placed on such joint action, much can be gained from a positive, realistic—above all, cooperative—approach to safety problems.

Cooperation is even more important where the problem area is double-ended: where both contractor and military personnel perform the same job and are subject to the same safety hazards. Therefore, in the following discussion of one such area—that of Bomarc transportation—any references to slip-ups on the military end of the airlift are meant to be strictly nonpartisan and objective. As long as there have been near accidents, it's better to use them as a guide for future safety than to pretend they never happened.

As this article goes to press, the safety record of Bomarc airlifts can be summed up in four words: so far, so good. You may recall, however, the optimist who jumped off the top of a New York office building. He was heard to yell the same thing as he passed the 20th floor: so far, so good.

This is not to imply—necessarily—that IM-99A on and offloading crews have been living on borrowed

Closeup of trailer shows hand brake linkage, towing cable. Handle forward of hand brake is steering selector. Two rings support missile in plane cradles.

**AEROSPACE SAFETY**

US Air Force

Popperfoto

---

**Joint Contracts.** In December 1961, a year after Thomas Pynchon's work as a technical writer at Boeing resulted in the publication of this article, his employers secured the prime contract for the first stage of the new Apollo Saturn-V rocket. The following year Pynchon left Boeing to finish assembling his own novel *V.*, published in 1963

**Towards a Common Center.** President Kennedy, *en route* to deliver his Rice University speech on 12 September 1962, is saluted by Houston residents. They welcomed not simply a President but also an influx of space program funding.

**Marriage of Convenience.**    John Glenn is welcomed back at Cape Canaveral by President Kennedy on 23 February 1962 following his successful oribital mission.

**Bright Angel.**    Christa McAuliffe celebrates her selection as the nation's first citizen astronaut during a parade through her home town of Concord, New Hampshire, on 20 July 1985.

**Shadowed Moment.**
Guest VIPs at Cape
Canaveral witness the
destruction of the space
shuttle *Challenger* on
28 January 1986.

Popperfoto

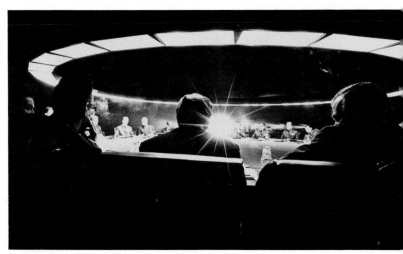

**Silent Frame.**   The War Room contemplates the way out in Stanley Kubrick's
*Doctor Strangelove.*

# 4

# Starship Troopers

## A National Job

> Thomas Carlyle in 1840 ... began to preach the idolatry of great men in his *Heroes and Hero-Worship*. For the growing sense of helplessness among the new urban masses he prescribed large doses of hero-worship. If you are a weak little man, get inside a big strong man. But we are not yet done with the long century of hero-worship or self-worship which began with the popular enthusiasm for Napoleon, 'the little corporal,' and which has seen the rise of the superman in theory, practice, and fantasy simultaneously.
>
> — Marshall McLuhan, *The Mechanical Bride* (1951)

> I have always felt especially close to the astronauts ... They represent the best this nation can produce; they are the folk heroes of our time.
>
> — Lyndon Johnson, *The Vantage Point* (1971)

John Kennedy had been investing in a space program, if not yet in spacemen for perhaps as much as three years by the time the Soviet Union launched the world's first manned capsule into orbit. Sputniks I and II carried more than four radio transmitters, a variety of scientific sensors, and a dog with them when they took off from Soviet Central Asia in late 1957: they also boosted the presidential prospects of a number of potential Democratic candidates, Kennedy included. In 1959 and 1960, as the Russians reproduced their initial successes with other space firsts, the issue of space conquest came to function as a key illustration of the unchallenged Soviet offensive in Kennedy's burgeoning campaign for nomination and election.[1]

Alongside the missile gap, Kennedy argued, the 'space gap' reflected the lack of national initiative and vitality under Eisenhower. 'With East and West competing to convince the new and undecided nations which way to turn, which wave was the future,' Sorensen later wrote, 'the dramatic Soviet achievements [in space] were helping to build a dangerous impression of unchallenged world leadership generally and scientific preeminence particularly.' Conversely, as Kennedy remarked during the campaign, Soviet primacy in space seemed to prove to the world that the United States had passed its 'high noon' and that 'the long, slow afternoon' of decline had already begun. Classified US Information Agency surveys leaked to the press ten days before the election confirmed that American prestige relative to that of the USSR had declined during the Eisenhower era and that in the eyes of her western allies the US was expected to remain behind the Soviet Union in space achievements for at least ten years. To dispel this impression Kennedy insisted that the nation had to prove itself on what he defined as 'our great New Frontier.' Since space was 'the symbol of the twentieth century,' the US could not afford to 'run second in this vital race.' 'To insure peace and freedom,' Kennedy told the voters a month before the election, 'we must be first.'[2]

The flight of Yuri Gagarin six months later only served to confirm that in 1961 the United States remained very much second in a two horse race. This was not, however, due to the effects of some dehistoricized, debilitating virus spread by the Republicans in the 1950s. Rather, initial Soviet supremacy resulted from a combination of material factors extending back well beyond the Eisenhower era. Unlike the United States, the Soviet Union did not have access to air bases within manned bomber range of its chief adversary. It therefore invested heavily in missile research and development from the end of World War II. As early as April 1947, Stalin had told the Politbureau that 'the creation of transatlantic rockets' was 'of extreme importance'

1. Michael Stoiko, *Soviet Rocketry: The First Decade of Achievement* (Newton Abbot: David & Charles, 1971), pp. 79-85, 88-9, 136-7; Evgeny Riabchikov, *Russians in Space*, trans. Guy Daniels, ed. Nikolai P. Kamanin (London: Weidenfeld & Nicolson, 1972), pp. 144-51, 153-5.

2. Theodore Sorensen, *Kennedy* (London: Hodder & Stoughton, 1965), p. 524; John Logsdon, *The Decision to Go to the Moon: Project Apollo and the National Interest* (Cambridge, Mass: MIT Press, 1970), pp. 64-6, 111; James Oberg, *Red Star in Orbit* (London: Harrap, 1981), p. 35; Vernon van Dyke, *Pride and Power: The Rationale of the Space Program* (Urbana, Ill.: University of Illinois Press, 1964), p. 23.

to the USSR since they could provide 'an effective strait jacket for that noisy shopkeeper, Harry Truman.'³ Thus whilst von Braun's rocket scientists labored in New Mexico and Texas, first under the restrictions of Truman's military economy drive in the second half of the 1940s and thereafter under the constraints imposed by Eisenhower's own preferential commitment to the Strategic Air Command, the Peenemünde engineers who had been shipped east in 1946 were exploited to the limits of Soviet capabilities. As a result, Soviet missiles were developed to carry the heavy and cumbersome nuclear warheads produced in the early 1950s whose delivery required very powerful launch vehicles; by contrast, American missile development was only given top priority by Washington in the mid-1950s, by which time advances in miniaturization and warhead design had rendered enormous boosters militarily obsolete. American rockets were not inferior as weapons, but the United States remained severely handicapped in terms of space flight.⁴

In the years after 1957 Soviet space achievements exploited that handicap to maximum advantage. As Khrushchev admitted in his memoirs, the rockets which launched the Sputniks were clumsy weapons, but their enormous lifting capacity made them ideal vehicles for Soviet propaganda. Consequently, Sputnik I was only the first in a series of worldwide advertisements designed to promote communism as the wave of the future and the USSR under Khrushchev as the vanguard of universal progress. Announcing the launch of the initial earth satellite in 1957, *Tass* had stressed 'how the freed and conscious labor of the people of the new socialist society' was turning 'even the most

---

3. William H. Schauer, *The Politics of Space* (New York: Holmes & Meier, 1976), pp. 8-14; Oberg, pp. 24-5; Stoiko, pp. 72-8; Walter A. McDougall, *The Heavens and the Earth* (New York: Basic Books, 1985), pp. 41-65; Zhores Medvedev, *Soviet Science* (Oxford: Oxford University Press, 1979), pp. 44-5. McDougall concludes that between 1945 and 1950, 'the Soviet budget for military research was several times greater than the American, and perhaps six times greater as a percentage of gross national product' (p. 51).

4. Clarence Lasby, *Project Paperclip: German Scientists and the Cold War* (New York: Atheneum, 1971), pp. 217-22, 248-50, 255-7; Frederick I. Ordway III and Mitchell R. Sharpe, *The Rocket Team* (London: Heinemann, 1979), pp. 318-43, 352, 371-3, 378-80; McDougall, pp. 54, 60, 77-8, 91, 97-103, 250-51; Schauer, pp. 122, 133-4; Jay Holmes, *America on the Moon: the Enterprise of the Sixties* (Philadelphia: J.P. Lippincott, 1962), p. 86; Charles C. Alexander, *Holding the Line: The Eisenhower Era, 1952-1961* (Bloomington: Indiana University Press, 1975), pp. 211-13; *New York Times*, 16 April 1961, sec. 4, p. 1.

daring of mankind's dreams into reality.'[5] Four years later Soviet flesh and blood in space was invested with even greater significance. Gagarin dedicated his flight on Moscow Radio 'to the people of a communist society, the society which our Soviet people are already entering and which, I am convinced, all the people on earth will enter,' while head of state Leonid Brezhnev stressed that the world's first cosmonaut had once been a foundry worker and that the nation's chief rocket designer was a former bricklayer: 'the lives of Korolev and Gagarin' were therefore 'a most vivid example of the wide prospects and opportunities that socialism opens up before the working man.' The simultaneous launch of two *Vostok* capsules in August 1962 was accompanied by statements stressing the Chuvash origins of cosmonaut Nikolaev whose flight illustrated how the Soviet Union had constructed a successful multiracial society. Likewise, the equality of women under socialism was demonstrated by the flight of Valentina Tereshkova less than a year later.[6]

Such propaganda was designed to vindicate not only Soviet society and technology but also Soviet history, particularly in the third world, which both Kennedy and Khrushchev viewed as the main battlefield in this latest phase of the Cold War. In 1917, the argument went, Russia had been an underdeveloped country under the control of international capitalism. Only after the Bolshevik Revolution had abolished private enterprise and introduced the planned economy had the Soviet Union been able to start building an advanced and just industrial society which would one day be capable of leading the way both on earth and in space. Where the United States had for years prided itself as the world's most advanced nation, Soviet space triumphs demon-

---

5. Nikita S. Khrushchev, *Khrushchev Remembers, Volume II: The Last Testament*, trans. and ed. Strobe Talbott (Harmondsworth: Penguin, 1977), pp. 70-71, 74-9, 86-90; Oberg, p. 33; Arnold Horelick, 'The Soviet Union and the Political Uses of Outer Space,' in Joseph M. Goldsen (ed.), *Outer Space in World Politics* (New York: Frederick A. Praeger, 1963), pp. 46-7, 49-50; *New York Times*, 16 October 1957, p. 43.

6. A. Romanov, *Spacecraft Designer: The Story of Sergei Korolev* (Moscow: Novosti Press Agency Publishing House, 1976), pp. 59, 65; Leonid Vladimirov, *The Russian Space Bluff* (New York: Dial Press, 1973), pp. 92-3; Schauer, pp. 104, 138; Frederick C. Barghoorn, *Soviet Foreign Propaganda* (Princeton: Princeton University Press, 1964), p. 201; Oberg, pp. 67-70; Oleg Penkovsky, *The Penkovsky Papers*, trans. P. Deriabin (London: Collins, 1965), p. 245; William Zimmerman, *Soviet Perspectives on International Relations 1956-67* (Princeton: Princeton University Press, 1969), pp. 165-79; McDougall, pp. 58-9.

strated that the Russian experience, not the American, was the blueprint for world development.[7] Such claims were pointedly emphasized when some of the central myths of the American historical experience were appropriated and applied to the Russians' own achievements. Thus Gagarin became the 'Columbus of the cosmos' whilst his successors were described as 'blazing a trail for all mankind' on the communist new frontier. The United States, by contrast, found itself dismissed as the backwash of history. In late 1957 the failure of Vanguard had allowed members of the Soviet UN delegation in New York City to ask their American counterparts if the United States would be interested in receiving aid under the Russsian technical assistance program for underdeveloped nations. Four years later the challenge issued by Khrushchev in a well-publicized telephone call to Gagarin on his return – 'Let the capitalist countries catch up with our country!' – was considered historically unanswerable. The flight was 'a confirmation of the correctness of Marxist–Leninist teachings.' 'Imperialism' was 'powerless to check the irresistible process of emancipation.'[8]

Kennedy's support for the Apollo project resulted from his desire, not so much to resist a development which he himself considered irresistible, as to reinstate the United States and its free enterprise system as the recognized instruments of liberation. On the eve of the launch of Sputnik I, Senator and candidate Kennedy had written that by moving 'outside the home fortress' the United States could 'challenge the enemy in fields of our own choosing.' Less than four years later the new President approved the Apollo project as a concrete expression of that challenge. For whereas Khrushchev had capitalized on the early Soviet lead in booster capacity, Kennedy chose an objective whose long-term nature provided sufficient scope for the application of superior American scientific, technical, industrial, and financial resources.

7. Frederick C. Barghoorn, *Soviet Foreign Propaganda* (Princeton: Princeton University Press, 1964), pp. 166-79; *New York Times*, 6 October 1957, p. 43, 13 April 1961, pp. 1, 14; V.P. Glushko, *Development of Rocketry and Space Technology in the USSR* (Moscow: Novesti Press Agency Publishing House, 1973), pp. 9-10; Schauer, p. 104. For a good example, see 'Editorial: Leading on Earth and in Space,' *International Affairs* (Moscow) 8, no. 9 (September, 1962), pp. 3-5.

8. *New York Times*, 16 April 1961, sec. 1, p. 50; Riabchikov, p. 32; Barghoorn, pp. 200-201; Glushko, p. 32; Oberg, p. 77; Logsdon, p. 101; Constance McLaughlin Green and Milton Lomask, *Vanguard: A History* (Washington: Smithsonian Institution Press, 1971), p. 210; Holmes, p. 136; McDougall, pp. 246-7.

Having asked Lyndon Johnson as Chairman of the Space Council to make an overall survey of American space capabilities and to provide recommendations for 'beating the Soviets' on 20 April 1961, the Vice-President's reply of 8 May confirmed what Kennedy had anticipated: the United States stood the greatest chance of victory by engaging in the lengthiest battle. The question the President had directed to White House and NASA staff two days after Gagarin's flight ('Is there any place we can catch them?') was thus answered in the affirmative: the Russians could be caught – and passed – en route to the moon.[9]

However, if the field of battle was chosen by Kennedy to suit the United States, the nature of the conflict and the strategies it demanded were agreed on by both sides. Khrushchev's investment in space and spacemen was matched by the American President from the outset. Where Moscow saw space achievement as a means of removing the unfavorable image gained on the streets of Budapest, Washington saw it as a way of repairing the damage inflicted at the Bay of Pigs. Where Khrushchev hoped that Sputnik and its successors would shatter conclusively the myth of western superiority, Kennedy believed that an American moon shot would recapture the frontier of scientific and technological achievement for the United States and with it the respect of the world. Where the Soviet Union played down the military dimensions of its space program, declaring it a peaceful endeavor designed 'to serve the welfare and happiness of all mankind,' the United States emphasized the division between NASA and Defense Department operations and vowed that American leadership in the space age would leave the heavens governed, 'not by a hostile flag of conquest, but by a banner of freedom and peace.'[10]

Within this common strategy the specific form of the American investment necessarily accommodated distinct national myths and political necessities. Both Kennedy and Johnson emphasized that the exploits of the astronauts placed them firmly in the American pioneering tradition. They went to the moon, the President told an audience at Rice University, Texas, in

9. Henry Fairlie, *The Kennedy Promise* (New York: Doubleday, 1973), p. 72; Logsdon, pp. 106, 109-26.

10. Holmes, p. 137; Schauer, pp. 41-3, 46, 62, 81, 92-105, 109-10, 136; Horelick, pp. 62, 64-6; John F. Kennedy, *The Burden and the Glory*, ed. Allan Nevins (New York: Harper & Row, 1964), p. 243.

September 1962, not because it was easy but because it was hard. Like their forefathers 'who tamed a broad continent and built the mightiest nation in the history of the world,' Johnson wrote in his memoirs, they 'blazed new trails across the untraveled wilderness of space.' They were in this sense agents of the American historical mission. But national destiny was not the only weighty object that Shepard and his successors were charged with. For in addition to the heroic efforts of the astronauts the execution of that mission required, according to both President and Vice-President, the voluntary and unified support of the American people.[11] Substantiating his inaugural rhetoric, Kennedy argued that in order to regain ascendancy in what he considered to be a race for global supremacy, Americans *en masse* would have to press themselves into the service of these modern-day pioneers. 'In a very real sense, it will not be one man going to the moon,' he told the Congress in May 1961, 'it will be an entire nation. For all of us must work to put him there.' The drive for leadership in space, Johnson announced a few months later, required nothing less than 'a fully cooperative, urgently motivated, all-out effort,' one which was and would remain 'a national job for all Americans, wherever they are assigned, wherever they live, whatever party they belong to.' In a way that was at once comparable to but distinct from that adopted by the Soviet Union, both American history and American society were invested in the space program generally and in the astronauts specifically. The men who pioneered on that final, inexhaustible frontier took with them more than just the hopes of NASA and the Democratic Party: they embodied a nation, a social system, a whole way of life. Their mission would make manifest America's destiny; their achievements would universalize the American Century.[12]

## From the Land of the Rubber Tubes

> They were looking for a certain type of animal who registered bingo on the meter.
>
> — Tom Wolfe, *The Right Stuff* (1979)

---

11. Kennedy, p. 244; Lyndon Baines Johnson, *The Vantage Point* (New York: Popular Library, 1971), pp. 285-6; Sorensen, pp. 525, 527-8; Logsdon, pp. 162-7, 170-71.

> Armstrong came in quickly ... Much like President Nixon or
> Wernher von Braun he would smile on command. Then a very
> useful smile appeared – the smile of an enterprising small-town
> boy. He could be an angel, he could be the town's devil. Who
> knew?

> — Norman Mailer, *Of A Fire on the Moon* (1971)

By 1962, Kennedy was including the astronauts, both collectively
and individually, within his own heroic vitalist elite. Many of the
astronauts had already placed themselves inside an identical
circle. Particularly in the early years, as they moved from the
realm of test pilot to that of NASA trainee and finally on to the
rank of proven astronaut, Kennedy's rising co-stars considered
themselves to be part of a select group who lived by higher standards of behavior than ordinary mortals, members of an enclosed
order united by shared qualities and common risks. They were
cool under pressure and skillful at the edge of disaster; they lived
by a cult of luck, calculating risks for status within a world of
permanent testing; they acted as bearers and protectors of those
all-absorbing, ostensibly supra-political American values of discipline and family, deity and flag. Sharing some intangible,
unspeakable, almost mystical faculty whose powers could only
adequately be revealed in a speeding car or a supersonic jet, they
belonged to what Tom Wolfe called 'the very Brotherhood of the
Right Stuff itself.'[13]

By electing to become astronauts, however, the members of
that brotherhood confronted a problem which had been evident
to a degree over thirty years earlier when Charles Lindbergh flew
the Atlantic in *The Spirit of St. Louis*. For if they, like their Soviet
counterparts, were bearers of societal and historical values, then
they were also – and more literally – bearers of American technology. Where the laurels pressed on Lindbergh as one of
nature's independent pioneers in 1927 had rested awkwardly on
the shoulders of a man whose achievement was the product of
engineering, organization, and accumulated experience, those
test pilots selected for NASA in 1959 who enjoyed the trappings

---

12. Logsdon, p. 128; *New York Times*, 14 October 1961, p. 6.

13. David Halberstam, *The Powers That Be* (London: Chatto & Windus,
1979), p. 385; Frank van Riper, *Glenn: The Astronaut Who Would be President* (New
York: Empire Books, 1983), pp. 35-7; Tom Wolfe, *The Right Stuff* (London:
Jonathan Cape, 1979), pp. 23-43.

of the 'right stuff' faced a more rigorous and systematic divestment.[14] As the men who preferred to stay airborne at Edwards Air Force Base were quick to point out, any volunteer moving from the established service branches to the new space agency would be exchanging a prestigious occupation for a 'Larry Lightbulb scheme.' He would no longer be the pilot of a ship but a pre-packed human cannonball. He would be converted, in Tom Wolfe's words, from a 'fighter jock' to 'a laboratory animal wired up from skull to rectum with medical sensors.' The division between the two was in practice less distinct than in this diptych. Nevertheless, the issue that concerns Tom Wolfe in *The Right Stuff* (and Norman Mailer also confronts in *Of A Fire on the Moon*) is what happens to heroic vitalism when its domain is subjected to the demands of instrumentation and remote command, when a flyer becomes 'Spam in a can.'[15]

What happens is what Géza Rószavölgyi dearly hopes will happen in *Gravity's Rainbow*, which is what Pynchon's power structure thrives on across and beyond the novel: the charismatic is progressively routinized. Just as in Mailer's fiction the basic tenets of the 'crazy old bastard' capitalist of the nineteenth century are superseded by those of the 'dull-eyed,' committee-based corporate capitalist of the twentieth, his mechanisms of laissez-faire and absolute surplus value by his replacement's circuitry of Fordism and relative surplus value, so in Wolfe's text the 'right stuff' is absorbed by the 'operational stuff.' In Wolfe's book the process is dramatically elaborated in the trajectory leading from Chuck Yeager, the epitome of the Edwards Air Force Base elite, to Wally Schirra, the penultimate and most proficient of the first seven Mercury astronauts. In Mailer's book it reaches its objective in the figure of Neil Armstrong, the first man on the moon. In either case, what the official historians of Project Mercury describe as 'machine-rating the men' can be handled in terms of two related developments: the conversion of

---

14. John W. Ward, 'The Meaning of Lindbergh's Flight,' *American Quarterly*, 10, no. 1 (1958), pp. 3-16.

15. Tom Wolfe, pp. 74-8. On the movement from pilot to astronaut, see Robert Jungk, *Tomorrow Is Already Here: Scenes From a Man-Made World*, trans. Marguerite Waldman (London: Rupert Hart-Davis, 1954), pp. 51-6; Roland Barthes, *Mythologies*, trans. Annette Lavers (1957; St. Albans: Paladin, 1973), pp. 71-3.

the pilots into redundant components, and their consequent revolt against redundancy (*GR* 81).[16]

The task of selecting a handful of Mercury trainees from the 110 servicemen who met the minimum requirements laid down by NASA and President Eisenhower reached its most concentrated phase during the spring of 1959, when three dozen volunteers were subjected to an exhaustive series of tests at the Lovelace Medical Research Clinic in Albuquerque, New Mexico, and the Wright Air Development Center Aeromedical Laboratory in Dayton, Ohio. At Lovelace they underwent over thirty different tests yielding chemical, encephalographic, cardiographic, and other data. There were x-rays and body radiation checks to endure, sperm samples and stool specimens to be supplied. A battery of 'straps, tubes, hoses, and needles' transmitting readings about a variety of organs from prostate gland and bowels to eyes, ears, nose and throat, helped doctors to quantify and evaluate every pilot's physiological profile and thereby to build up 'more complete medical histories ... than probably had ever before been attempted on human beings.' At Wright Field the candidates, by now instructed to identify themselves by number only, were put through other, equally taxing investigations. On the one hand there were a variety of physical endurance tests measuring the trainees' ability to withstand heat and acceleration, high energy sound and vibration. On the other there were thirteen different psychological probes, ranging from a 'complex behavior stress test' (which was quickly dubbed the 'idiot box') and thematic apperception studies to psychiatric interviews (in which candidates quickly learned to phrase all answers in the language of the 'operational stuff') and experiments in sensory deprivation.[17]

The sole marine to be selected for the Mercury flights, Lieutenant Colonel John H. Glenn, Jr., found the warnings of men like Yeager fully justified. At Lovelace the volunteers 'were treated as little more

16. Tom Wolfe, pp. 46-68, 379-86, 416; Norman Mailer, *Of A Fire on the Moon* (Boston; 1970; New York: New American Library, 1971), pp. 26, 32-7, 40, 165; Kees van der Pijl, *The Making of an Atlantic Ruling Class* (London: Verso, 1984), pp. 1-34; Loyd S. Swenson, Jr., James M. Grimwood, and Charles C. Alexander, *This New Ocean: A History of Project Mercury*, NASA SP-4201 (Washington, D.C.: National Aeronautics and Space Administration, 1966), p. 223.

17. Wolfe, pp. 88-97; Swenson, *et al.*, pp. 160-62; van Riper, pp. 130-36; Joseph N. Bell, *Seven Into Space: The Story of the Mercury Astronauts* (London: Ebury Press, 1960), pp. 56-63.

than specimens, to be prodded and probed with maddening frequency.' Selection was based on physical and technical criteria rather than immeasurable heroic qualities. As Wolfe puts it, 'the kind of stuff you were made of as a *pilot* didn't count for a goddamned thing.' If you won the competition it would not be in the air but 'on the examination table in the land of the rubber tubes.'[18] Moreover, as other pilots had also anticipated, this was only the logical consequence of the strictly circumscribed functions proposed for the astronaut inside his capsule. At best he was to be a test bed for biosensors, a 'sponge of information' (Glenn's description), or a redundant component prepared to fill the gaps where technology failed. At worst his primary task would be to do nothing under stress. The testing at Wright Field was designed, in the words of USAF psychiatrist George Ruff, to discover those candidates 'with no evidence of impulsivity, who will refrain from action when inactivity is appropriate.' Subsequent training was intended to adapt out any remaining anxieties they might harbor. Whilst proposals for tranquillizing astronauts to ensure unimpeded performances were ruled out, their restricted operational domain provided little scope for the exercise of any 'righteous stuff.' In the opinion of some NASA consultants, at least, it might prove dysfunctional; for them, according to Tom Wolfe, 'an experienced zombie would do fine.'[19]

Not surprisingly, the prospective astronauts had other ideas. Within days of entering the Lovelace clinic a number began resisting what they considered an undignified and unnecessary regime. The decision of naval Lieutenant Pete Conrad to deliver his first stool specimen wrapped in a red ribbon helped exclude him from the Mercury team (a decade later he walked on the moon), but it won the admiration of his fellow trainees and – along with other, less emphatic protests – effectively alerted clinicians to their opinions. Since those responsible for Project Mercury inside NASA's Space Task Group knew that the astronauts were likely to become the agency's chief salesmen, they proved sensitive to the trainees' requirements and reduced the number of medical experiments accordingly.[20] This sensitivity

18. van Riper, p. 130; Wolfe, p. 95; Gene Farmer and Dora Jane Hamblin, *First on the Moon* (London: Michael Joseph, 1970), p. 105.

19. Wolfe, pp. 179-82; van Riper, pp. 132-5, 160-62; Swenson, *et al.*, pp. 174, 177, 194-5.

20. Wolfe, pp. 91-5, 191; van Riper, pp. 131-2.

necessarily extended to the outside world. Within weeks of the initial unveiling of NASA's public faces in April 1959, STG officials were also emphasizing that, far from being mere passengers, the astronauts 'were to prove their full potential as pilots' in space. Thus encouraged, the chosen seven began working to reorient both the hardware and software of spaceflight to their own designs. Alterations requested and incorporated included a proper window in place of a single periscope and a pilot-operated hatch instead of an externally sealed one. Likewise, following the successful sub-orbital missions of Alan Shepard and Virgil Grissom, 'capsules' were renamed 'spacecraft' in the vocabulary of Mercury staff. Such changes had at least some operational justification. As chief Flight Director Christopher Craft, Jr., later remarked, after the first manned flights had demonstrated how well astronauts could perform in space, earlier assumptions about the necessity of automation could be revised. Man might therefore remain a component in the circuits of spaceflight, but he need no longer be redundant. In the words of human engineering expert Edward Jones, the astronaut of the future would prove to be 'an essential component who can add considerably to systems effectiveness when he is given adequate instruments, controls, and is trained.'[21]

This integration of the astronaut, which Jones likened to an evolutionary process, accelerated rapidly over the next decade. In 1960 the Mercury capsule incorporated 40,000 functioning parts, included 120 instruments and controls for the occupant to monitor or operate, and required checks on 130 items in the course of a normal flight. By 1969 the Command Module which took Armstrong, Aldrin, and Collins to the moon contained 2 million functioning parts, carried 650 'switches, dials, meters, circuit breakers, controls and displays' for the commander alone, and programmed thousands of actions for all three crew members during its eight-day mission. Surrounded by systems and subsystems for environmental control, guidance and navigation, communications, and a dozen other essentials requiring their attention, the Apollo XI crew were anything but redundant.[22] And yet, in endeavoring to make room for the 'right stuff,' that crew's predecessors had in practice only paved the way for the

---

21. Swenson, *et al.*, pp. 174, 177, 194-5, 247, 487; Wolfe, pp. 191-3.
22. Swenson, *et al.*, pp. 179, 195, 245-6; Mailer, *Fire on the Moon*, pp. 163-5, 212-19; *New York Times*, 17 July 1969, p. 38.

'new breed' of elite figures of whom Neil Armstrong was by 1969 the exemplar. The commander of Apollo XI was no longer just an expert pilot but also a skilled engineer, mechanic, and technician. Far from being an independent pioneer he was, in Norman Mailer's words, 'the representative of a collective will.' His expertise, already shared with perhaps ten other astronauts, was linked to that of hundreds of ground-based consultants; his language, 'full of anisotropic functions and multiple encounter trajectories,' derived not from the world of gambling but from the consoles of MIT; his vital functions were constantly monitored in Houston and his every significant action rehearsed and guided by on-board and earth-based computers. In *Of A Fire on the Moon* Mailer suggests that the flight of Apollo XI made 'heroism's previous relation to romance' appear 'highly improper.' By the same token, the elevation of Armstrong, Aldrin, and Collins sealed the 'operational stuff's most fitting and sober victory.'[23]

The victory belonged not only to the crew of Apollo XI and the 'operational stuff' but, by the nature of their production, to Pynchon's entire Operation. Kennedy's overstatement notwithstanding, a lot of people – and things – *had* worked to put man on the moon. Indeed the flight was a celebration of the Rocket State which produced it. The spacecraft constituted a microcosm of incipient totalitarian life. In 1961 Alan Shepard's Mercury capsule had seemed 'like an extremely compact modern kitchen ... with all the gadgets running at once.' Eight years later the larger Apollo Command Module incorporated an even more comprehensive selection of post-war produce – from IBM computers to headache pills, from color television cameras to sleeping pills, from freeze-dried food to stomach pills – which led Norman Mailer to describe it as a combination of sixteen different environments: at once laboratory and bathroom, television studio and bedroom, kitchen and gymnasium.[24] Beyond this, and more specifically,the spacecraft constituted a comforting affirmation of incipient totalitarian ease. If the United States in the summer of 1969 seemed a hostile environment (which for a

23. Wolfe, p. 416; Mailer, *Fire on the Moon*, pp. 40, 49, 101, 220, 222, 226, 227, 261, 315; Peter Ryan, *The Invasion of the Moon 1957-1970*, rev. and enlarged edn (Harmondsworth: Penguin, 1971), pp. 77-8; *New York Times*, 17 July 1969, pp. 31, 36, 39.

24. Wolfe, p. 257; Mailer, *Fire on the Moon*, p. 214; Ryan, pp. 49, 75, 77; *New York Times*, 22 July 1969, p. 27.

number of reasons it did), then the Apollo XI capsule looked reassuringly like the haven of security to which young Ilse Pökler had aspired in *Gravity's Rainbow*: a 'shelter in time of disaster,' a 'Happyville' well beyond 'Pain City.' It was, indeed, the ultimate post-war desirable residence. It was automated, transistorized, and fully furnished; it placed everything within easy reach; and it came packed with disposable produce (aside from their scientific instruments and the flag, the astronauts dumped about $1 million worth of used equipment on the moon). Finally, and as if to prove the point, it accommodated on completion a 'high society' almost without equal: not only more exclusive than Bing Crosby's in Newport but also beyond the reach of both Russians and rebels.[25]

The occupants had, of course, worked hard to get where they were. The spacecraft was a microcosm of incipient totalitarian life and its abundance, and the astronauts were condensed testimonies to the rewards of incipient totalitarian labor. They had been raised by the disciplines of hot war and tested on the American Century's cutting edges (all but one of the Mercury astronauts served in World War II and/or Korea, whilst both Aldrin and Armstrong saw action over Korea); they had converted their imperial qualifications into a place on a large high technology enterprise's post-military stepladder; and then they had worked their way up.[26] In turn the routine conditions of that work bore a distinct resemblance to those experienced by Mossmoon and millions of other employees. At Wright Field the first astronauts endured a regime whose objectives derived from Taylor's *Principles of Scientific Management* and whose methods had much in common with those presented in Orwell's *Nineteen Eighty-four* (surveillance) and Packard's *Hidden Persuaders* (motivational research).[27] In subsequent years all their successors spent thousands of hours on a variety of procedures trainers, rehearsing

---

25. *New York Times*, 22 July 1969, p. 26. The Command Module actually carried part of the *High Society* cast in the shape of taped Frank Sinatra music. See Farmer and Hamblin, p. 103.

26. Bell, pp. 66-83; Ryan, pp. 56-7; Wolfe, pp. 41-3.

27. The 'complex behavior stress test' (or 'idiot box') at Wright Field bore an uncanny resemblance to 'The Switchboard,' one of 'several forms of disciplinary procedure' introduced by Doctor Benway of the Freeland Republic in William Burroughs contemporaneous *The Naked Lunch*. As Wolfe notes, it 'appeared to be not only a test of reaction times but of perseverance or ability to cope with frustration.' William Burroughs, *The Naked Lunch* (1959, London: Corgi, 1974), pp. 39, 42; Wolfe, p. 97.

flight plans repeatedly in order to be able to perform their tasks efficiently, automatically, and without fear – like the servant figures programmed in Asimov's *I, Robot,* the brainwashed pods cultivated in *Invasion of the Body Snatchers,* or 'The Complete All-American De-anxietized Man' unveiled by Dr. Schafer in William Burroughs' *The Naked Lunch.* As Mailer notes, even en route to the moon the astronauts faced a 'variety of repetitive chores and duties. The thousand subdetails of routine rocket housekeeping were upon them,' from charging batteries to changing aerials.[28]

But the routine chores and duties of home and workplace, while they contributed to the astronauts' status as representatives of American life and labor, were themselves hardly sufficient to accommodate the heroic vitalist faith. The Operation in this sense required more than just their discrete operational stuff. Manned spaceflight was, however, anything but discrete. For whereas the Edwards Air Force Base pilots had constituted a self-styled and essentially unpublicized elite (when Yeager first broke the sound barrier in October 1947, the US Air Force kept it secret for eight months), the astronauts were from the outset presented by NASA as figures for public consumption. And it was as leading men, as stars, that they retained their elite status. Millions of Americans prepared dinner, checked lists, and adjusted television aerials daily, but only a handful did so in front of an audience composed of the same millions: the paying customers installed in the Orpheus Theater's integrated network of 'invisible rooms.' While the right stuff was absorbed by the operational stuff, the Operation dispersed the heroic vitalist faith via the Rocket State's 'peep-show machines.' Like John F. Kennedy, the astronauts were publicly launched well before their official platforms were hammered out (*GR* 450).[29]

## To the Land of the Vacuum Tubes

No movie queen ever got more attention.

— Virgil Grissom, *Gemini!* (1968)

28. Swenson, *et al.,* pp. 234, 240-45, 343; Wolfe, pp. 164, 182-5, 232, 248-9; Brian Ash (ed.), *The Visual Encyclopedia of Science Fiction* (London: Pan, 1977), pp. 161, 176; Burroughs, p. 124; Mailer, *Fire on the Moon,* p. 220.

29. Wolfe, pp. 59-62.

The elevation of the astronauts to heroic status began on 9 April 1959, with their presentation before the nation's media inside NASA's temporary Washington headquarters. It completed its first stage that August when they signed an authorized deal with *Life* which gave Luce's magazine exclusive rights to their stories and each of the trainees an equal share of $500,000 over a three-year period along with a veto over all material bound for publication. Thereafter the seven men found themselves despatched on what one described as a 'year-long dog and pony show' involving personal appearances at NASA construction sites and contractor's facilities, personal appearances before Congress, and personal appearances before the cameras.[30]

They had entered the star system, and that, as Edgar Morin emphasizes, was first of all a system of production which defined them as capital to be invested in or raw material to be enriched. In common with other publicity campaigns, directors, technicians, and backroom staff were employed to provide a combination of cameras and wardrobe, lighting and make-up, cutting and close-ups capable of converting dolls into idols (decked out in suits made of very 'Peculiar Polymer') and adding luster to valuable property. More than thirty years earlier Charles Lindbergh had found himself reproduced overnight as a time-honored national type; now the astronauts discovered that this still common investment strategy was to be applied to them in turn. Within days of their presentation the *New York Times* anticipated Kennedy's own approach by describing the Mercury trainees as men of courage and conviction who 'spoke of "duty" and "faith" and "country" like Walt Whitman's pioneers.' Five months later *Life*, having previously boosted both Sputniks and Kennedy, became a 'pacesetter for the flattery' of the astronauts by publishing the first two in a series of cover stories on the manned space program in which its leading men became all-American heroes and their leading ladies all-American heroines: the men patriotic and brave, the women devoted and respectable (*GR* 699).[31]

---

30. Wolfe, pp. 126, 140-42; van Riper, pp. 136-42, 146-9; Swenson, *et al.*, p. 438.

31. Edgar Morin, *The Stars*, trans. Richard Howard (New York: Grove Press, 1960), pp. 40-46, 54-5, 135-6; van Riper, p. 146; Wolfe, pp. 122-4, 141-2; *Life*, 14 September 1959, 21 September 1959; *New York Times*, 12 April 1959, sec. 4, p. 8; Ward, p. 6. On publicity for the Apollo program, see Hugo Young, Bryan Silcock, and Peter Dunne, *Journey to Tranquillity* (London: Jonathan Cape, 1969), p. 54.

Yet the production of public icons depended on shadows as well as spotlights. For the wives this meant that their *Life* cover photographs were retouched and that a curtain was drawn over such things as Annie Glenn's stutter and Marjorie Slayton's previous marriage. For their husbands it meant that remarks like those made by Chuck Yeager dismissing Project Mercury as work suitable for monkeys received minimal coverage, and that behavior deemed unworthy (such as extra-marital sex or Gordon Cooper's separation) was flushed, in Tom Wolfe's allusion to *Nineteen Eighty-four*, 'down the memory hole.' As John Glenn's biographer later wrote, 'anything that was not standard Apple Pie America simply did not see print.' At Edwards Air Force Base a private taboo once nourished the 'right stuff'; now public silence sustained it across the land.[32]

In Orwell's novel, however, the memory holes consumed not merely inconvenient evidence but the raw material of history itself. Over a decade later that capacity remained intact. For in censoring the behavior of the astronauts, *Life* and others constrained recognition of, amongst other things, the strains imposed on military marriages by limited incomes, frequent relocation, extended separation, and the pressure of work. They likewise suppressed consideration of the backstage maneuvering for selection by NASA and of the competition for the right to fly first. Discussion of such issues as the power structure of the military and of American society as a whole was consequently made to appear superfluous and irrelevant.[33] In an earlier analysis of the content of popular magazine biographies, Leo Lowenthal recorded that in such texts:

> the vocational set-up of the dramatis personae is organized as if the social production process were either completely exterminated or tacitly understood and needed no further interpretation ... The real battlefield of history [therefore] recedes from view or becomes a stock backdrop while society disintegrates into an amorphous crowd of consumers.[34]

---

32. Wolfe, pp. 23-38, 122, 127-8, 157-60; van Riper, pp. 146-51.

33. George Orwell, *Nineteen Eighty-four* (1949; Harmondsworth: Penguin, 1954), pp. 30-38, 63, 126-7; Wolfe, pp. 154, 163-77, 214-27; van Riper, pp. 103, 109-11, 117-24; Young, *et al.*, p. 170.

34. Leo Lowenthal, 'The Triumph of Mass Idols,' in *Literature, Popular Culture, and Society* (Englewood Cliffs, N.J.: Prentice-Hall, 1961), pp. 116, 123.

In the late 1950s these observations remained pertinent. In *Life* history came rigged up as a second nature, identity was portrayed as part of an apolitical drama of initiative and fortune, and class relations were absorbed as status within community. Restated in the language of Roland Barthes, *Life* presented a semiological structure of values disguised as an inductive structure of self-evident facts. In the terms provided by Jacques Ellul, it propagated sociological, integrative, and rational propaganda.[35]

Once inducted by the star system, the astronaut went on to become something more than a 'kind of frozen capital,' a commercial asset, and a carefully prepared luxury item, however. Abstracted in the course of his production from those very means of production (like Benjamin's 'orchid in the land of technology'), each of *Life*'s front men completed his elevation in the form on which the Rocket State most conspicuously thrived. For if the astronaut was trained as a component he was simultaneously traded as a commodity: in Morin's words, as a mass-produced piece of merchandise with 'all the virtues of a standard product adapted to the world market, like chewing gum, refrigerators, soap, [or] razor blades.'[36] As such his value should have been subject to approximation by the laws of supply and demand, and in an immediate sense it was. On the one hand supply was self-evidently limited: by the time Neil Armstrong set foot on the moon only twenty-two other Americans had flown into space by ballistic capsule. On the other hand demand was clearly both widespread and heavy: as both Wolfe and Mailer record, the astronauts' appeal extended across much of the class structure, from the Convair plant workers to the Washington Congressmen who celebrated the Mercury pilots, and from the trades unionists, southern mill workers, and 'families of poor Okies' who were drawn to Cape Kennedy in July 1969 to the cross-section of the power elite who combined in the dignitaries' stand. However, such observations could not easily be evaluated within the calculus of capitalism. Even as the astronauts flooded the commercial

---

35. Barthes, p. 131; Jacques Ellul, *Propaganda: The Formation of Men's Attitudes*, trans. Konrad Keller and Jean Lerner (1968; New York: Vintage, 1973), pp. 62-8, 74-6, 84-7. See also Guy Debord, *Society of the Spectacle* (Detroit: Black & Red, 1977), paras 24, 72.

36. Morin, pp. 136-9; Walter Benjamin, 'The Work of Art in the Age of Mechanical Reproduction,' in *Illuminations*, ed. and intro. Hannah Arendt, trans. Harry Zohn (London: Jonathan Cape, 1970), pp. 230-31, 233, 235; Susan Buck-Morss, *The Origins of Negative Dialectics* (Hassocks: Harvester, 1977), pp. 146-9.

circuits of the Orpheus Theater, their profit margins remained at once evident and undefined.[37]

The reason for this binary condition was directly connected to the nature of the Rocket State's development. The astronauts may have been as devoid of price tags as the celebrities Lowenthal had earlier studied, but this was not so much because they lacked a market as because their market had become so dispersed. Anyone who purchased *Life*, who bought petrol to drive to the launch site, who then stayed in a Florida hotel, who got a new television for the space specials, who rode the subway to a post-flight parade, or who picked up the souvenirs of success was spending in order to stare at the astronauts or to share in the spectacle which starred them. Anyone who did any of these and comparable things was simultaneously furthering the integration and absorption of the Rocket State's elevated expenditure structure.[38] Such behavior thereby revealed NASA's leading men in their finished state as polymorphically attractive commodities. At the same time it transformed them, by virtue of their sheer ubiquity, from discrete commodities into a continuous context: the groundbase of McLuhan's propaganda. Their pricelessness, in turn, became more than just a sign of the accountant's limited reach; it was a token envincing the boundless and indistinct nature of the no man's land and the invisibility the Operation had sought since Gerhardt von Göll had first begun screening *New Dope* (*GR* 527, 745).[39]

The argument has another dimension. For if the inability of a purely material analysis to accommodate the astronaut resulted, in one sense, from the nature of the Rocket State's structure of expenditure, in another it was a direct consequence of its structure of production. The astronaut, even as he was undergoing preparation in procedures trainers and photographers' lenses, made his appearance on a larger, more public stage: at that point where the work of the projectors' hidden hands reached the audience's 'hushed and invisible rooms' – on the third parabola

---

37. Ryan, pp. 199-201; Wolfe, pp. 126-7, 148, 238-9, 268-70, 388; Mailer, *Fire on the Moon*, pp. 59-61, 84-5. See also *New York Times*, 16 July 1969, p. 22, 17 July 1969, p. 21, 20 July 1969, sec. 4, p. 1.

38. Lowenthal, p. 131. For some idea of one sector of expenditure, see the discussion of publishing and Apollo in the *New York Times*, 16 July 1969, p. 30. On the ubiquity of media, see Umberto Eco, *Travels in Hyperreality*, trans. William Weaver (London: Pan, 1987), pp. 148-50.

39. Burroughs, p, 9.

or the 'nonstop revue' of Pynchon's *Gravity's Rainbow* (*GR* 4, 681).[40] He was, in other words, like all of Morin's stars,

> made from a substance compounded of life and dream ... [and thus possessed of] a syncretic personality in which the real person [could not] be distinguished from the person fabricated by the dream factories and the person invented by the spectator.[41]

Since the astronaut's means of production included his consumers, the analysis must also consider the star system's induction of the Rocket State's paying customers: the descendants of Greta Erdmann, Miklos Thanatz, and Tyrone Slothrop. For as Jacques Ellul repeatedly states, behind all effective propaganda lie not only propagandists who produce it but also propagandees who need it.[42]

## The Ambivalent Sacred

> Parsons, his attention caught by the trumpet call, sat listening with a sort of gaping solemnity, a sort of edified boredom. He could not follow the figures, but he was aware that they were in some way a cause for satisfaction ... 'The Ministry of Plenty's certainly done a good job this year,' he said with a knowing shake of his head.
>
> — George Orwell, *Nineteen Eighty-four* (1949)

> There's plenty of emotion going on, but you don't hear it.
>
> — Mrs. Joan Aldrin (16 July 1969)

In the face of an authoritative statement announcing increases in the production of everything from food to clothing and from books to babies, Tom Parsons' inability to follow the Ministry of Plenty's figures mattered little: he embraced the Party line at the formal level without regard to its content or context. The figures presented to the American public by NASA in April 1959, and later by the Kennedy and Johnson administrations, remained equally elusive. Even after a decade of manned spaceflight, Norman Mailer could characterize a typical working-class

---

40. See above, pp. 83-4.
41. Morin, p. 105.
42. Ellul, *Propaganda*, pp. 37, 121, 138-60.

witness at Cape Kennedy in July 1969 as 'uncertain whether to cry from pride or the all-out ache that he does not really comprehend the new machinery.'[43] Nevertheless, he, like Parsons, was in attendance, absorbed by the immensity of it all, awash with nationalist admiration and, along with nearly a million others, eager for the show to begin.

The explanation for such common assent lay precisely in the fact that both Orwell's functionary and Mailer's mill worker erased or sought to evade the contexts of their desires. Thus in *Nineteen Eighty-four* Parsons and many others embraced the Party's fictional standards of security, confidence, and promise as life-lines in the midst of a drab, under-provisioned, crowded, polluted, dangerous, and propagandized environment. Likewise, in Mailer's text the worker's awed attendance was grounded in that sense of anonymous inferiority born of economic and social constraint characteristic of the late capitalist societies.[44] As Clive Bush's analysis of the iconography and sociohistorical context of George Washington in *The Dream of Reason* reveals, such phenomena were hardly novel but remained entirely characteristic of a society supposedly founded upon a revulsion against idolatry and kingship: the sheer quantity of energy expended since the Republic's earliest days on the drama of public heroism in the United States indicated a 'clamour for social identity' which was and remains 'deeply allied to thwarted social needs.'[45]

If the bases of what Orrin Klapp terms 'celebrity watching' were the Rocket State's mobilizing contradictions – the capitalist system's demand for consumption and its simultaneous restrictions on expenditure, its celebration of self-reliant individualism and its need for a dependent and mass-produced labor force, its appreciation of security and its multiplication, exaggeration, and dissemination of liabilities – then the essence of the hero's appeal

---

43. Orwell, pp. 48-52; Mailer, *Fire on the Moon*, pp. 59-61.

44. Orwell, pp. 238-9; Ellul, pp. 149-51. The contexts of the astronauts' elevation, given contemporary liberal expression in J.K. Galbraith's *The Affluent Society* (Boston: Houghton Mifflin, 1958), were more rigorously analyzed in, for example, Gabriel Kolko, *Wealth and Power in America* (New York: Praeger, 1962), Murray Bookchin, *Our Synthetic Environment* (New York: Alfred A. Knopf, 1962), and Michael Harrington, *The Other America* (New York: Macmillan, 1962). Bookchin published under the pen-name Lewis Herber.

45. Clive Bush, *The Dream of Reason* (London: Edward Arnold, 1977), pp. 19-45; Norman O. Brown, *Love's Body* (New York: Vintage, 1966), p. 114. See also Morin, p. 98.

was his appearance as the agent and image of their resolution. As Jacques Ellul points out, to the extent that any propagandized society systematically undermines friendship, trust, independence, and security and systematically fosters feelings of frustration, guilt, inferiority, and the desire for power, a hero, leader, or celebrity will appear effortlessly to satisfy those needs. In a damaging parody of compact theory as presented by Hobbes, Montesquieu, and others, he is elected by an ill-defined and unsigned but completely binding contract to provide the appearance of security in exchange for the material consequences of impotence and alienation. Like both his rocket and capsule, therefore, the astronaut becomes another 'shelter in time of disaster,' more or less accommodating depending on his specific domain and construction; in Norman O. Brown's words, 'a safety deposit bank' which would prove no more secure than any of the Rocket State's other privatized vantage points.[46]

The terms of the contract which NASA first drew up in April 1959 elevated the astronauts to headline status in the Orpheus Theater and installed the general public as its satisfied customers. This articulation did not, however, inaugurate a political society with access to the stage of history quite as Brown argues in *Love's Body*. Rather, the elevation of NASA's public faces lifted the curtains on an imitative drama in which the historical contexts of both astronauts and audience (as employees and commodities, as products and consumers) were recast in the form of a depoliticized ritual of immutable nature. The astronaut was a representative man, a public figure, and an 'artificial person' as men like Kennedy and Johnson implied and as his polymorphic production confirmed, yet he was simultaneously taken up as a charismatic hero whose election was at best virtual and as a sacred deity whose position was a matter of faith. Correspondingly, even as the public were, as citizens, his formal (if indirect) electors and, as taxpayers, his investors, then they were also, as pacified spectators, his disenfranchised devotees installed in his secular sepulcher. In such conditions the society articulated by the manned space program was anything but the political, historicized body of Brown's abstracts. The nature of political representation was already impossible to define in the mid-eighteenth

---

46. Orrin Klapp, *Collective Search For Identity* (New York: Holt, Rinehart, & Winston, 1969), p. 241; Ellul, pp. 147-60, 171-5; Bush, 20-22; Brown, *Love's Body*, p. 117.

century; by the mid-twentieth the term alone had become meaningless and therefore ubiquitous only as propaganda. The astronauts were 'representative men' (to recall Emerson) in a society whose struggle for effective representation and accountability had long since been lost.[47]

As in *Destination Moon*, the astronaut's appeal lay less in his abstract constitution than in his evident reliability and his practical application. To an audience strung out across the Rocket State's structure of interlocking contradictions he appeared, like his co-star in the White House, to combine the public and private dimensions of civil society in a drama of sanctioned and unconstrained life. In the public sphere the astronaut was presented as one of the most celebrated of Kennedy's 'watchmen on the walls of world freedom.'[48] Whether in public parades, press conferences, or photo calls, he embodied those universal American values of piety and hard work, of family and flag. Whether in military service or atop Army and Air Force launchers, he 'maintained a sense of discipline while civilians abandoned themselves to hedonism [and] a sense of honor while civilians lived by opportunism and greed.' As amplified by a mass media intent on conveying what Tom Wolfe described as 'the *proper emotion*, the *seemly sentiment*, [and] the *fitting moral tone*,' his ostensible willingness to risk death in an heroic endeavor for the most noble ideals and his appearances as an efficient, well-adjusted, modest, and restrained citizen ensured that, like George Washington before him, he was elevated to the status of a Carlylean archetype: at once 'a social reassurance of America's public character' and a 'morally instructive parable of state.'[49]

Simultaneously the astronaut appeared in his personal capacity to have everything his audience were denied and to be everything that they could not be. For those citizens who felt trapped inside what Edgar Morin describes as 'the mean and anonymous life' of the stalls and who wanted to enlarge themselves 'to the dimensions of life in the movies,' he fulfilled continuously thwarted desires.[50] Cummings, stranded on a remote island in *The Naked and the Dead*, had dreamed of being promoted from the 'tedium and routine, [the] regulations and

47. Brown, pp. 109-17; Bush, pp. 22-5; Barthes, pp. 72-3.
48. Kennedy, p. 277.
49. Wolfe, pp. 39-40, 114-22; Mailer, *Fire on the Moon*, pp. 24-37; Lowenthal, pp. 128-9; Orrin Klapp, *Symbolic Leaders* (Chicago: Aldine, 1964), p. 44; Bush, pp. 32, 41-6.

procedure' of warfare and had found in the scream of artillery a sense of power to match his own 'deep, boundless ambition.' The astronaut, by contrast, shot promptly to the top of a structure whose 'manmade thunder and light' eclipsed the Army's most powerful shells.[51] Similarly, whereas Slothrop, having run the gauntlet at Potsdam in *Gravity's Rainbow*, had dreamed of being signed up by Truman's 'showbiz types' for a life of glamor and fame, the astronaut drew the spotlight as the star of one of the greatest of presidential shows. Like Charles Lindbergh, described by one correspondent in the *American Magazine* in 1927 as 'our dream of what we really and truly want to be,' the astronaut embodied what Guy Debord called 'the inaccessible result of social labor' under capitalism, dramatizing its by-products in the form of a model life 'magically projected above' the crowds. Particularly during the early years of manned spaceflight, when politicians and journalists put their names on the map, but also in the late 1960s, when the impending moon shot made an astronaut Hollywood society's most sought after dinner guest, Morin's description of the movie stars' appeal applied equally to NASA's leading men. Like their celluloid partners they came to 'satisfy the gossip columns of the heart.' Like Kennedy himself, they revitalized the American heartland.[52]

As both the terms of their production and consumption indicated, the astronauts' idealized operation depended on a concomitant process of obliteration. They functioned, that is to say, not only as an embodiment of the American Dream but also as a vehicle for the discharge of its underlying nightmares: to borrow Leo Lowenthal's phrase, as 'a readily grasped empire of refuge *and* escape,' simultaneously fusing and detonating the social system that produced them. This composite performance, which fitted the needs of Pynchon's elevator network precisely, was evident in the reactions of the crowds attracted to the countdowns. Thus the woman at Cape Kennedy who told the *New York*

---

50. Morin, p. 98. See also, Klapp, *Collective Search*, pp. 211-14; Lowenthal, pp. 116, 123; Morin, pp. 59-60, 176-9; Theodor Adorno, 'Freudian Theory and the Pattern of Fascist Propaganda,' in Andrew Arato and Eike Gebhardt (eds), *The Essential Frankfurt School Reader* (New York: Urizen Books, 1978), pp. 125-6; Ellul, pp. 171-2.

51. Norman Mailer, *The Naked and the Dead* (1948; St. Albans: Panther, 1964), pp. 476-82.

52. Ward, p. 6; Debord, para. 60; *New York Times*, 20 July 1969, p. 39; Morin, pp. 6, 20.

*Times* 'it's so thrilling, maybe the engine will explode' revealed as much about the motives of the audience who crowded the Florida coastline in July 1969 as those who whispered 'God bless them, God bless them' at lift-off.[53] To a public whose appetite for Mickey Spillane and his counterparts showed no sign of diminishing (as Kurt Vonnegut wrote a few days before the launch of Apollo XI, 'what they like are shows where people get killed'), the astronauts and their mission constituted only the latest installment in a drama of violation upon which social security was taken to depend. They partook of what Georges Bataille called 'the ambivalent sacred,' a binary star system in which, like every other god in Morin's stellar liturgy, they were at once consecrated as redemptive guardian angels and offered up as sacrificial lambs.[54]

To the extent that the space program's leading lights constituted an 'empire of refuge and escape,' their sovereignty was vested in the ritual of launching itself. For it was in the act of launching that the chorus first sung by Coover's commuters in *The Public Burning* and subsequently swelled and echoed across the stalls of the Orpheus Theater received its most dramatic articulation. As Mailer's description of the Apollo XI lift-off indicates, it was at once an apocalyptic climax and a draining catharsis, a 'fury of sound' combining a thrilling sensation of power with an overwhelming feeling of helplessness, a 'ball of fire' providing excitement and relief for that cross-section of philobats and ocnophiles distributed within its domain.[55] Redefined in Bataille's language, it constituted an act of *dépense* which seemed to release the inhabitants of the Orpheus Theater's 'invisible

---

53. Lowenthal, p. 135 (emphasis added); *New York Times*, 17 July 1969, pp. 21, 22. See also Mailer, *Fire on the Moon*, pp. 36-7, 80-81.

54. Kurt Vonnegut, 'Excelsior! We're Going to the Moon! Excelsior!' *New York Times Magazine*, 13 July 1969, p. 10; Lewis Mumford in the *New York Times*, 21 July 1969, p. 6; Tom Wolfe, pp. 126-7, 238-9, 269-70; Michelle Richman, *Reading Georges Bataille: Beyond the Gift* (Baltimore: Johns Hopkins University Press, 1982), p. 47; Morin, pp. 86, 90, 98; Henri Hubert and Marcel Mauss, *Sacrifice: Its Nature and Function*, trans. W.D. Halls (1898; London: Cohen & West, 1964), pp. 98-9. See also Norman O. Brown's formulation in *Love's Body*, (New York: Vintage, 1966), pp. 118-19), and Eric Bentley's description of heroic vitalist figures as both authoritarians and rebels in *The Cult of the Superman* (London: Robert Hale, 1947), p. 238.

55. Mailer, *Fire on the Moon*, pp. 90-94, 184; Michael Balint, *Thrills and Regressions* (London: Hogarth Press/Institute of Psychoanalysis, 1959), pp. 23-31; Robert Coover, *The Public Burning* (1977; Harmondsworth: Penguin, 1978), p. 247.

rooms' from the constraints and complaints of their lives: the endless cycles of labor and consumption; the inescapable burdens of credit cards and credit charges, mortgages and televisions, phone calls and Oldsmobiles; and the insoluble anxieties of life in the lonely crowd. To the all-consuming offspring of Greta Erdmann and the docilely multiplying descendants of Miklos Thanatz, as to the joyless progeny of General Cummings and Dominus Blicero and the myriad other would-be *Führers* dispersed across the darkened stalls, the lift-off appeared to recreate what Claude Lévi-Strauss described as 'that fleeting moment when it was permissible to believe that the law of exchange could be evaded, that one could gain without losing, enjoy without sharing.' It proffered an escape valve at the very heart of the joint command structure; a way out which released time and space itself from the binding terms of the Rocket State's contract (*GR* 464, 487-8, 758).[56]

But the evacuation sought by the audience drawn within the shadow of the Apollo spacecraft and its predecessors was as hallucinatory as that pursued by the crowd of passengers caged in beneath the screaming, glass, and girders of *Gravity's Rainbow*, as immaterial as the 'promise [or] prophecy of Escape' contained inside the 'deep cry of combustion' which accompanied the launch of Blicero's rocket from the clearing on Luneburg Heath. For even as it appeared to stimulate and absorb its witnesses, the lift-off simultaneously affirmed their material status as powerless onlookers whose sense of involvement derived from a vicarious combination of identification and detachment and whose participation was little more than voyeuristic. In observing the rocket's thunderous ascension the audience were collectively stunned into silence. Like Ahab's crew, petrified before their own 'iron statue' in *Moby Dick*, they found themselves rendered speechless by the Rocket State's greatest public monument: spellbound by its technological wizardry, paralyzed by its 'nightmare of sound,' turned to stone within its 'vast burning of air,' and, like their fictional

---

56. Claude Lévi-Strauss, *The Elementary Structures of Kinship*, rev. ed, trans. James Harle Bell, John Richard von Sturms, and Rodney Needham (London: Eyre & Spottiswoode, 1969), p. 497; Richman, pp. 3, 30-31, 81; Norman Mailer, *Naked and the Dead*, p. 478; Georges Bataille, *Visions of Excess: Selected Writings, 1927-1939*, ed. Allan Stoekl, trans. Allan Stoekl, Carl Lovitt, and Donald Leslie, Jr. (Manchester: Manchester University Press, 1985), pp. 116-30; Norman O. Brown, *Life Against Death* (1959; Middletown, Conn.: Wesleyan University Press, 1970), pp. 266-8, 274-7.

predecessors, 'rooted to the deck [with] all their eyes upcast.' If, during the first seconds of flight, Norman Mailer experienced 'a poor moment of vertigo at the thought that man now had something with which to speak to God,' then it would not be the mill worker or the Operation's other employees speaking (Neil Armstrong's job was every bit as exclusive as Bing Crosby's or Grace Kelly's) but the powers that directed their labor. Those who accepted an astronaut as the nation's perfect father substitute simultaneously installed themselves as his dependent and impotent children, surrendered their sovereignty to a 'desexualized cult object,' and, within a drama of sacrificial redemption, indulged in an exercise in masochistic abandon. They became, not partners in national rejuvenation but the reusable Gottfrieds that Kennedy once sought: in Norman O. Brown's words, 'spectator[s] thrilled at [their] own execution (*GR* 3, 758).'[57]

## Strung Into the Apollonian Dream

The largest crowd ever to witness a space launching surged along beaches, parks, roads, and highways tonight and waited excitedly for the Apollo XI flight to the moon tomorrow ... Nearly a million tourists are expected to pack Brevard County for the launching. The wildly assorted group of visitors includes former President Lyndon B. Johnson, members of the poor people's campaign, African and Asian diplomats, youths carrying Confederate flags, vacationing families, hippies, scientists and surfers, and students and salesmen ... Everywhere there are buttons and bumper stickers that read 'I Was At The Apollo XI.'

— *New York Times* (16 July 1969)

These people have a right to demonstrate as long as they don't break the law and tie up traffic.

— Cocoa Beach Sheriff Leigh Wilson (14 July 1969)

The space spectaculars which culminated in the voyage of Apollo

---

57. *New York Times*, 17 July 1969, p. 21; Mailer, *Fire on the Moon*, p. 93; Brown, *Love's Body*, pp. 67-71, 119-25; Adorno, pp. 121-7, Hubert and Mauss, p. 95; Debord, para. 25; Herman Melville, *Moby Dick; or The Whale*, ed. Harold Beaver (Harmondsworth: Penguin, 1972), pp. 594, 614-15, 617-18, 643-4; Bush, pp. 43, 45, 56-7.

XI in July 1969 proved to be absorbing and elaborate fabrications. Even as they appeared to take place above and beyond the Rocket State, celebrating its operation but at the same time maintaining their autonomy, these latest installments of Gerhardt von Göll's *New Dope* remained both integral products and archetypal expressions of a social system whose essential cleavages – like the separation of power and the division of labor – were more concrete. As three distinct but intersecting extrapolations from Pynchon's novel will demonstrate, they constituted the ultimate vehicles for that 'progressive *knotting into*' which had initially been pursued as the 'way out' during late 1944, when an earlier generation of overburdened evacuees in tightly packed carriages began moving across the first rocket-trained city at the beginning of *Gravity's Rainbow* (*GR* 3, 171-3).[58]

First, that the flight of Armstrong, Aldrin, and Collins marked the culmination of a process of integration and evacuation whose origins lay amidst the ruins of Pynchon's Oven State could be illustrated by a series of structural diptychs. In 1945 the launching of Gottfried completed the precarious integration of Clive Mossmoon, the basic personnel unit of the Rocket State's postwar reproduction; twenty-five years later the launching of men as enamored of rockets as Gottfried tenuously united millions of Mossmoon's mobilized offspring. In 1945 Tyrone Slothrop became a 'Rocketman' in order to gain the benefits of what was a potentially hazardous but profitably intoxicating experience at Potsdam; in 1969 nearly a million 'consistent personae' dispersed by his decay turned out as boosters for NASA with the same end in mind at Cocoa Beach. During the final weeks of World War II the Rocket State was simultaneously inaugurated at an improvised firing-point and at an improvised conference center 'lit up like a Hollywood premiere'; a generation and more later it reached its apotheosis at the Saturn V's custom-built launch pad upon which Norman Mailer saw trained 'giant arc lights, as voluminous in candlepower as the lights for an old-fashioned Hollywood premiere.' Finally, in 1945 the origins of the Rocket State's synthetic arcs were surrounded by military men like Blicero, by politicians like Harry Truman, and by 'showbiz types'

---

58. Debord, paras 7, 8, 23, 24, 72; Young, *et al.*, pp. 15-16, 144; Khachig Tölölyan, 'War as Background in *Gravity's Rainbow*,' in Charles Clerc (ed.), *Approaches to Gravity's Rainbow* (Columbus: Ohio State University Press, 1983), p. 33.

like Mickey Rooney; twenty-five years on their joint apex was besieged by a celebrated company of thousands, from General Westmoreland and Johnny Carson to Lyndon Johnson and Cardinal Cooke.[59]

The Rocket State's increased capacity in turn evinced the extent to which the articulating joint of the Oven State, the *Führer* principle, had been reintegrated, reground, and redistributed following its stiffening and seizure in 1945. Thus, where Blicero had been forced to abandon the public sphere in favor of his isolated colonial dream world, to escape what he described as 'this cycle of infection and death' by dissolving the flux of history into a frozen autobiographical frame, his offspring were now easily taken in by what was an absorbing social occasion, their desire 'to break out' contained by their experience of blast-off. Where Blicero, cornered in 'his own space,' could only dream of 'discovering the edge of the world,' of finding 'that there *is* an end,' the products of his scattered seed were now to witness man's first arrival on that 'new Edge.' Blicero had slammed shut the doors of the 00000's 'oven'; his descendants now found themselves encouraged to look at the interiors of Apollo XI, labelled and lit up for all. As the remarks of those installed in the Orpheus Theater demonstrated, the flight to the moon constituted the realization of President Kennedy's voluntary totalitarian dream: an audience of small-time *Führers* passively finding their collective identity via willful subordination within an elaborate drama of state. In the words of a border patrolman watching on television in California: 'I was there, I was part of it, I saw it happen.' In the words of a foreign student vacationing at the Cape: 'I am part of history, I have seen the launch.' (*GR* 98-9, 486, 670, 722-4)[60]

Finally, this absorption satisfied, to varying degrees, the main requirements of those post-war planning initiatives articulated by Géza Rószavölgyi and Gerhardt von Göll in *Gravity's Rainbow* and thereby testified to the survival and extension of the Operation beyond what the former described as 'the awful interface of V–E Day.' On the one hand, while the decade-long primary mission of the American space agency continued to rely on personalities for its propagation against Rószavölgyi's advice, what appeared to be

59. Mailer, *Fire on the Moon*, p. 57; *New York Times*, 16 July 1969, p. 22.
60. *New York Times*, 17 July 1969, p. 1, 20 July 1969, sec. 4, p. 1, 16 July 1969, p. 22.

its 'powerful leader[s]' remained directed within the 'abstractions of power' he preferred and dependent on what was clearly a 'powerful program' for their influence. Charisma, defined by Rószavölgyi in 1945 as 'a terrible disease,' had to this extent been routinized and the *Führer* principle defused. On the other hand, the manned space program quite literally made von Göll's day, combining dramatic illusion with substantive action and science fiction with science fact, creating an enormous object of attraction inside the no man's land's 'nonstop revue,' and projecting what Pynchon describes as '[a]nother world laid down on the present one and to all appearances no different' – a dehistoricized and depoliticized *theatrum mundi* in which the Rocket State had completely absorbed its surrounding Zone, Debord's spectacle had finally dominated social life, and incipient totalitarian society had become an enclosed maze of interlocking passages; a homeostatic and terminal system, that is, with no other goal but itself. (*GR* 80-1, 527, 664, 681)[61]

Under such conditions what had appeared to elevate the Orpheus Theater's customers to a higher plane only left them standing in the stalls. Like previous features in von Göll's long-running *New Dope*, the manned space program proved to be an exercise in stasis which articulated a mechanism of feedback within the Rocket State rather than one of delivery from it. The consequences for the viewers were no different from those for previous patrons. Since, to borrow Walter Benjamin's formulation, 'the audience's identification with the actor [was] really an identification with the camera,' they found themselves taken up as no more than transmitting mechanisms within the Rocket State's machinery.[62] Thus qualities of individual security and liberty ostensibly fostered within the astronaut's 'empire of refuge and escape' were recast in their more characteristic capitalist forms as isolation, entrapment, and dependence, whilst the cumulative debts absorbed by NASA's leading 'safety deposit bank[s]' were rescheduled rather than repaid and redistributed rather than written off. In turn such transactions ensured that, rather than rejuvenating a harmonious community in which both private and public interests and factional and general needs found mystical accord, the diverse operations which collectively made up the American space industry actually affirmed and

---

61. Debord, paras 3, 4, 18, 31.
62. Benjamin, p. 230.

extended the domain of Clive Mossmoon's more coercively molded masses. By carrying out what Guy Debord describes as 'a controlled reintegration of workers' which 'reunite[d] the separate but reunite[d them] *as* separate,' they fostered neither trust nor social responsibility but the mutual alienation characteristic of the lonely crowd's *ersatz* community (*GR* 100).[63]

Notwithstanding the residual angst voiced by the *New York Times* the day after the moon landing, the achievements of Armstrong and his crew carried no irony when measured against the shortcomings of the society they claimed to represent.[64] Nothing had gone wrong, either on board Apollo XI or inside the space program's more extensive public works. On the contrary, schedules had been kept to the minute. Thus at the point of its ostensible completion the contract between the astronauts and their formal constituents was revealed to be neither broken nor bent but, in keeping with its initial specifications, thoroughly prejudicial. From the beginning, the selection of NASA's public faces as representative men had been carried out, not in the public domain but behind interlocking screens of bureaucracy, science, and marketing. It had not so much been determined by an active electorate as endorsed by a disabled audience. It had depended, not on their considered judgement as critical citizens but on their undivided attention as loyal consumers. To the extent that the space program's heroes inaugurated a political society and gave the public what they believed to be access to history, they did so inside a theater of operations where history had been restaged as second nature and politics recast as administered behavior. As instruments of spectacular propaganda, the nature of their representation proved less virtuous than apparent and more virtual than effective.[65]

However, if the final words inscribed on the commemorative plaque attached to the Lunar Excursion Module on Apollo XI

63. Debord, paras 29, 172; Ellul, pp. 174-5; H.T. Wilson, *The American Ideology: Science, Technology and Organization as Modes of Rationality in Advanced Industrial Societies*, (London: Routledge & Kegan Paul, 1977), pp. 184, 187, 189; Brown, *Life Against Death*, p. 269.

64. 'For all his resplendent glory as he steps forth on another planet, man is still a pathetic creature, able to master outer space and yet unable to control his inner self,' *New York Times*, 21 July 1969, p. 16.

65. Ellul, p. 180; Debord, paras 12, 25; Adorno, pp. 124, 178 n. 11; Gustave LeBon quoted in C. Wright Mills, *The Power Elite* (1956; New York: Oxford University Press, 1959), p. 87; Brown, *Love's Body*, p. 112.

('We came in peace for all mankind') asserted a motive whose authenticity remained unsubstantiated, then the advertising copy splashed across the pages of the *New York Times* during the week of the flight ('Sony promises you the moon,' 'RCA – Shooting for the Moon since '58,' 'Brillo offers you the Moon. Free,' and 'Time-Life: *To the Moon* – $19.95') confirmed which constituents the astronauts did speak for, and why.[66] In the late 1950s, when Norman Mailer noted that '[t]hrough the post-war years prosperity [sic] has been maintained in America by invading the wage-earner in his home,' the first astronauts were completing their selection procedures. Ten years later, when John O'Neill restated and contextualized Mailer's essential point, their successors were preparing for flights to the moon. As their technical capabilities extended so did their commercial capacity. With American business interests still 'faced with the problem of deficient consumption structurally related to the class distribution of income,' and therefore still finding it 'necessary to invade the psychic space of workers and consumers through raising levels of expectation or through deepening levels of credit,' so the Apollo crew were appropriated as advertising hoardings for anything from scouring pads to swimming pools, just as the Mercury trainees had been used to sell *Life* a decade before. They became, that is, second cousins to the promotions men of Hollywood and Madison Avenue analyzed by Marshall McLuhan in *The Mechanical Bride*: representatives of organizations 'constantly striving to enter and control the unconscious minds of a vast public ... in order to exploit them for profit.' As Norman Mailer put it in *Of A Fire on the Moon*: 'NASA was vending space [and] Armstrong was working directly for his corporate mill.'[67]

## Beyond the Monkeypod Life

Glenn seemed to eat this stuff up. He couldn't get enough grins or

---

66. Ryan, p. 126; *New York Times*, 15 July 1969, pp. 2, 46, 20 July 1969, p. 21, 21 July 1969, p. 13.

67. Norman Mailer, 'From Surplus Value to the Mass-Media,' in *Advertisements for Myself* (New York: 1959; London: Panther/Granada, 1968), p. 355; John O'Neill, 'Public and Private Space,' in Trevor Lloyd and Jack McLeod (eds), *Agenda 1970: Proposals for a Creative Politics* (Toronto: Toronto University Press, 1968), p. 84; Marshall McLuhan, *The Mechanical Bride: Folklore of Industrial Man* (1951; Boston: Beacon Press, 1967), p. 97; Mailer, *Fire on the Moon*, p. 45. See also George Bataille's formulation in *Visions of Excess*, p. 156.

handshakes, and he had a few words filed away in every pocket. He would even come back to Langley and write cards to workers he had met on the assembly line, giving them little 'attaboys,' as if they were all in this thing together, partners in the great adventure, and he, the astronaut, would never forget his, the welding inspector's, beaming mug ... Oddly enough, it seemed to work.

—Tom Wolfe, *The Right Stuff* (1979)

Marketed during the 1960s as the representatives of all Americans and ultimately of all mankind; promoted as agents of liberation from unspecified burdening conventions of the past; yet produced and deployed in the interests of a community of specific concerns whose powers depended on the maintenance of established and equally specific conventions (such as the rights of private property) and whose conception of liberty remained correspondingly narrow – the astronauts proved to be instruments of liberalism, the Rocket State's characteristically expansive mechanism.[68] Their appeal to that state's other leading man was therefore only to be expected: as John F. Kennedy's promotion of their exploits confirmed, they constituted appropriate vehicles for his own distinctive promise. The successful Democratic candidate of 1960 campaigned under the banner of a 'New Frontier' and promised a break with the past only to fill his cabinet with a cross-section of the nation's industrial, financial, and legal elite; the astronauts correspondingly epitomized the imaginative life of the new frontiersman and articulated a thrilling fiction of boundary breaking only to return to the security of the earth's more established divisions. Kennedy as President presented himself as the principled advocate of the nation's moral obligations; his chosen co-stars were promoted as the dutiful and dedicated expressions of American public honor. If in Orrin Klapp's typology both politician and astronauts performed as 'seductive heroes,' presenting a semblance of liberation for those experienced in constraint with the characteristically vibrant stasis of so many liberal characters, then in Richard Maltby's sense they appeared as equally enchanting method actors: Kennedy offering an image of unconventionality as a substitute for political radicalism after Marlon Brando (a comparison made by Mailer and others), and the astronauts acting as professional warriors

---

68. See above, pp. 88-90.

combining skill, composure, and technology in the face of imminent martyrdom to defend the endangered customs of domesticated life, like the stars of *The Magnificent Seven*.[69]

But however transient and immaterial the benefits provided to their audience were, the leading men themselves found their credits paying off in more material ways. Just as Yul Brynner and Steve McQueen rode off the set of *The Magnificent Seven* to greater personal fortune, so their technologically more advanced contemporaries in the Mercury program came back to earth as proven performers whose fame and popularity could also be harnessed for private gain. Indeed, even before they completed their roles, NASA's trainees began to enjoy the fruits of the Operation's investment in space. On signing the *Life* contract in August 1959, M. Scott Carpenter exchanged his basic naval lieutenant's income of $7,200 per annum, which he supplemented with housing subsidies and flight pay, for a guaranteed $24,000 per annum from the publishing contract alone. His six colleagues, none of whom earned more than $12,000 per annum, received the same reward, a sum whose importance became all the greater when NASA restricted its trainees' flying opportunities and therefore their previously routine flight pay of $145 per month. When Alan Shepard became the first American in space, however, the *Life* payments were revealed as only preliminary returns – advances for the astronauts and insurance policies for their families – from what was clearly a much larger contract. 'Installed as a national hero on the order of a Lindbergh' on his successful recovery, Shepard found himself deluged, not only by the cheers of the crowds in New York and Washington, but also with endorsement proposals worth some $500,000, the equivalent

69. Fairlie, pp. 125-7; van der Pijl, pp. 196-9; Bruce Miroff, *Pragmatic Illusions: The Presidential Politics of John F. Kennedy* (New York: David McKay, 1976), pp. 7-10; G. William Domhoff, 'Who Made American Foreign Policy 1945-1963?' in David Horowitz (ed.), *Corporations and the Cold War* (New York: Monthly Review Press, 1969), pp. 45-6; Klapp, *Collective Search*, pp. 217-27; Richard Maltby, *Harmless Entertainment: Hollywood and the Ideology of Consensus* (Metuchen, N.J.: Scarecrow Press, 1983), pp. 225-54, 261-78; John Sturges, dir., *The Magnificent Seven*, with Yul Brynner and Steve McQueen, United Artists, 1960. On Kennedy as a method actor, see Norman Mailer, *The Presidential Papers* (1963; St. Albans: Panther, 1976), p. 61; Peter Collier and David Horowitz, *The Kennedys: An American Drama* (1984; London: Pan, 1985), pp. 289-92. As the latter wrote of Kennedy, apparently without irony: 'in contrast to the radicalism of the left and of the right, his was a radicalism of the status quo.'

of the total *Life* deal the seven men had signed less than two years before.[70]

Shepard's experience was itself overshadowed some nine months later when John Glenn became the first American to orbit the planet. Unlike the six other astronauts, Glenn had already established a degree of popular recognition before the completion of the *Life* contract. Credited with six distinguished flying crosses and seventeen air medals following combat in both the Pacific and Korean wars and subsequent service achievements, his establishment of a new coast to coast flying speed record for the US Navy in July 1957 had brought him favorable television and radio coverage, enthusiastic press reports (including a 'Man in the News' profile in the *New York Times*), and successful appearances on two television quiz shows. Subsequently, his self-promotion at the press conference introducing NASA's seven trainees in April 1959 had made him the astronauts' most public and publicized face.[71] When he was selected to become the third American in space, however, the specific context of his mission ensured that its successful completion would elevate him beyond his already established position to a stature unrivalled in the American pantheon of space. For whilst the sub-orbital flights of Shepard and Virgil Grissom in May and July 1961 had temporarily restored national pride in the wake of Gagarin's coup, those achievements had in turn been eclipsed by the day-long, seventeen-orbit flight of Gherman Titov in August 1961.[72] Glenn was a celebrity before launch, but his successful recovery would turn him into the object of a rarely equalled campaign of adulation, one intended to counter the renewed claims of the Russians in space and also to overwhelm the obstacles erected by the Russians across Berlin in the weeks after Titov's return.

Deification proceeded apace. Encouraged by the rapidly expanding NASA publicity machine, some 50,000 spectators gathered along with hundreds of reporters at Cape Canaveral on 20 February 1962 to watch Glenn's launch whilst another 100

---

70. Tom Wolfe, pp. 100, 141, 151, 268-73; van Riper, p. 146; Swenson, *et al.*, pp. 237-8. Shepard later became a moon-walking multi-millionaire. See Young, *et al.*, pp. 171-3.

71. Bell, pp. 66-8; van Riper, pp. 25, 103, 112-17; Philip N. Pierce and Karl Schuon, *John H. Glenn: Astronaut* (New York: Ballantine Books, 1962), pp. 21-34, 40-54; *New York Times*, 17 July 1957, pp. 1, 5; Wolfe, pp. 114-20.

72. Tom Wolfe, pp. 297-8, 307-8; Swenson, *et al.*, pp. 377-9; *New York Times*, 13 August 1961, pp. 1, 32.

million followed the countdown on television. Less than five hours later, after a three-orbit mission covering some 75,000 statute miles, NASA's latest star returned as history incarnate. Glenn's first footsteps on board the recovery ship *USS Noa* were marked out by white paint for exhibition at the Smithsonian Institution. In Washington the US Post Office issued the orders which put on sale special postage stamps commemorating the event. Back on dry land Glenn was submerged by a flood of congratulatory telegrams including one from Charles Lindbergh himself. Meanwhile, after post-flight analysis at the Cape the much travelled capsule commenced its fourth orbit of the planet: a seventeen-nation world tour which would display *Friendship Seven* to millions more people before finally bringing it back to rest a year later at the Smithsonian, next to the Wright Brothers' *Flyer 1*, Lindbergh's *Spirit of St. Louis* – and Glenn's own distinguished footprints.[73]

Having completed two days of post-flight examinations and recovery at Grand Turk Island in the Bahamas, the capsule's former occupant returned to Patrick Air Force Base, Florida, on 23 February. Surrounded by hundreds of reporters and Lyndon Johnson, the latter 'oozing protocol all over' Glenn's wife while simultaneously 'straining to ... pour Texas all over him,' Glenn was welcomed by the Vice-President as 'a great pioneer of history' and by sixth grade pupils at Cocoa Beach Elementary School as 'America's new Columbus.' Following a fifteen-mile motorcade which attracted over 100,000 'wildly cheering' spectators, he was then greeted by the President himself at Cape Canaveral and presented with NASA's Distinguished Service Medal 'for exceptionally meritorious service to the Government of the United States.'[74] Three days later Glenn flew to Washington for a parade which drew 250,000 people despite the pouring rain, and for a carefully rehearsed address to a Joint Session of Congress which drew applause both thunderous and tearful. At the White House he accepted the keys to the city and placed his signature on the National Geographical Society's 'flyers and explorers globe' next to the names of Byrd, Lindbergh, Hilary, and others. Lastly, on 1 March 1962 Glenn (still closely accompanied by the Vice-President), along with his family and the other astronauts,

---

73. Swenson, *et al.*, pp. 419-26, 436, 640-41; van Riper, pp. 95-6, 176; Pierce and Schuon, p. 132.

74. Pierce and Schuon, pp. 122-8; Wolfe, p. 344; van Riper, pp. 34-5.

received the largest ticker-tape welcome in New York City's history from a crowd estimated at around four million who jammed the entire eighteen-mile motorcade route.[75]

Shepard and Glenn led the astronauts *en masse* away from the bleak and poorly appointed military installations of the test pilot's life and towards the higher circles of the Orpheus Theater; in Tom Wolfe's words, away from the 'Monkeypod Life' of long hours, limited incomes, and constant insecurity and into 'the upper reaches of American protocol,' that 'strange world where Those People, the people who do things and run things, actually exist[ed.]' After Shepard's flight the seven Mercury astronauts travelled to the White House to meet both President and Mrs. Kennedy; following Glenn's mission they not only took their wives to the Executive mansion but also found themselves installed in suites at the Waldorf–Astoria, given a celebration dinner at the Tower Suite on top of the Time–Life building by Henry Luce himself, and entertained by his top executives with the city's most appropriate Broadway play: *How to Succeed in Business Without Really Trying.*[76]

But if such treatment constituted the occasionally rich icing on an already substantial cake to some of the Mercury team, to John Glenn it afforded access to a new career divorced from the routine procedures of NASA. According to his biographer, as early as the 'Project Bullet' coast to coast record run of July 1957, Glenn had discovered how much he enjoyed 'basking in the public eye.' Thereafter he had *campaigned* for election as an astronaut, calling in favors and tapping his old boy network as effectively as a Lyndon Johnson, and having successfully taken his seat had begun yet another campaign for selection as the first man to fly. NASA's decision to choose the primary pilot by peer group vote may have undermined his entire electoral strategy, but his involvement in the public relations side of the job nevertheless stimulated his desire to move still higher inside the world of the 'engineered yes.'[77]

If his Congressional address of April 1962 witnessed John

75. Pierce and Schuon, pp. 132-6; Wolfe, pp. 345-50; van Riper, pp. 40-46; Swenson, *et al.*, p. 437. Compare to the receptions accorded Gagarin and Titov noted by Riabchikov, pp. 37-46, 170.

76. Wolfe, pp. 38, 100-102, 273-5, 348-52, 358-62.

77. van Riper, pp. 45-9, 103, 117, 121-4, 143-4, 151-4; Wolfe, pp. 134-9, 148, 216-19; Vance Packard, *The Hidden Persuaders* (1957; Harmondsworth: Penguin, 1962), pp. 177-84.

Glenn starting to change his means of ascent, it was partly because he had gained the support of a powerful new booster in the President of the United States. The popular reception accorded Alan Shepard the previous May had convinced John Kennedy of the potential electoral benefits of space in general and of spacemen in particular, and had led him to conceive of the latter as the ultimate expressions of his New Frontier spirit, in contrast to his predecessor who looked upon them as little more than experimental volunteers. After their achievements and new assignments had helped rescue the President's own political fortunes following the flight of Yuri Gagarin and the fiasco at the Bay of Pigs, Kennedy moved to promote them accordingly. It was an act fusing business with pleasure. As his former Georgetown neighbor and *Washington Post* editor Benjamin Bradlee recalled: 'Kennedy identified enthusiastically with the astronauts, the glamor surrounding them, and the courage and skill it [took] to do their job.' The emotional bond which fertilized the resulting marriage of convenience would, the President hoped, enable him to raise a nationwide family of depoliticized Democratic voters.[78]

Kennedy and Glenn were first introduced at the White House on 5 February 1962, before the latter went south for his flight. Comparable in age and both veterans of the Pacific War, the two men shared in a world of risk taking, manliness, high technology, and 'cool,' and their relationship blossomed quickly following the astronaut's successful return. Inserted into the new orbit of Kennedy family life at Hyannisport, Glenn and his wife sailed with the President on board *Honey Fitz*, played tennis and touch football, and generally exceeded the Kennedys themselves in the realm of physical activity. But if NASA's most celebrated hero continued to enjoy his exertions on earth, his aerial exercises were soon to be terminated. Kennedy had invested much political capital in Glenn and looked on him as a potentially valuable ally in his forthcoming battle for re-election in 1964. On completion of the Mercury program in June 1963, the President ordered his new protégé grounded. Like Yuri Gagarin in the Soviet Union, Glenn was deemed to have become an important national asset: whatever thoughts he still harbored about additional ventures in space, his political mentor considered him too valuable a commodity to be risked in action again. Instead he

---

78. Wolfe, pp. 270-76; Logsdon, pp. 100-112, 121-9; Benjamin Bradlee, *Conversations With Kennedy* (New York: Norton, 1975), p. 191; Halberstam, p. 385.

was to be transferred to the hold of another elaborate machine, the Democratic Party. For his part, the potential candidate needed little persuading. Restricted to a desk job and ceremonial duties for NASA and strategic Congressmen, at 42 beyond the original age limit for astronauts, and already considering his own political career, Glenn informed NASA Administrator James Webb that he would no longer be available for the space agency's publicity operations. From now on he would appear for himself.[79]

During the early autumn of 1963, the transition seemed to near completion when Glenn accepted the offer of a Kennedy-assisted entrée into the Senate as Democratic candidate for his home state of Ohio, a crucial contested region in the forthcoming national elections. A few weeks later all such arrangements collapsed. Kennedy was dead and the new President's political needs precluded a John Glenn candidacy in Ohio. Whether the assassination bolstered his desire to run for high office or nearly persuaded him to abandon the fight, Glenn evidently learned one important lesson from the death of his political guardian. Money remained the lifeblood of politics. Kennedy had had a bloodline to Kennedy, Sr.; Glenn would have to become his own Lyle Bland.[80]

But just as Alan Shepard's successful mission had attracted endorsement proposals worth $500,000, so Glenn's orbital flight brought in its wake a deluge of even greater commercial offers which, unlike the sacks of fan mail that remained piled up at the Library of Congress twenty years after his exploit, promised the very financial resources a renewed political career required and therefore demanded prompt attention. Presented with a variety of invitations, the largest of which guaranteed him $1 million for his ten-year endorsement of a General Mills breakfast cereal, Glenn eschewed the world of advertising in favor of a directorship at Royal Crown Cola, seats on several other corporate boards, and the accumulation of stock options concentrated in Ohio. These in turn led to a corporate vice-presidency and presidency and a career in international business during the mid-1960s. At the same time Glenn continued to cultivate the space-derived interests he had established in Florida as early as 1959, when he and

---

79. van Riper, pp. 35-9, 49-56; Wolfe, pp. 362, 393-4.

80. van Riper, pp. 56-7, 184. Glenn eventually withdrew from the race for the Democratic nomination following a domestic accident. In Lyndon Johnson's memoirs the man he once cultivated relentlessly became an unperson.

some of his Mercury colleagues invested their *Life* savings in Henri Landwirth's Cape Colony Inn at Cocoa Beach. Following a profitable decade of involvement in the region's NASA-fuelled expansion, Glenn and Landwirth used the former's corporate status to secure the backing of a Cleveland bank in order to build and operate a Holiday Inn franchised hotel near both Disney World in Orlando and a Cape Kennedy complex then engaged in the main phase of the Apollo launchings. The 'Holiday Inn East of Disney World' made Glenn a millionaire, allowing him to extend his field of operations to include other Holiday Inns, real estate, and some seventy-five additional companies, thereby providing the financial basis for a successful Ohio senatorial campaign in 1974 and a presidential nomination campaign in 1984, by which time his net value exceeded $6 million.[81]

Rising from a small town childhood through distinguished military service and national heroism on to personal wealth, political influence, and presidential aspiration; recovering from setbacks, disadvantages, and disappointments through hard work and personal initiative, John Glenn epitomized the fiction of the self-made American: a small-time Lyle Bland whose personal experiences confirmed the promises of freedom his 1962 flight had been commissioned to extend.[82] And yet in the process of elevation he exposed such freedom as a set of interlocking traps within a characteristically Faustian Rocket State. Produced, like his mentor in the White House, by that state's mechanisms of engineering and propaganda, and advanced, like Kennedy, within its capitalist precepts, Glenn soon found himself equally dependent and enclosed. Thus when the President hypocritically questioned the propriety of the *Life* contract and other such commercial arrangements during the summer of 1962, NASA's leading man had to lobby fiercely for its retention since, as Tom Wolfe put it, the astronauts 'had begun to look upon the *Life* deal in the same way they looked upon the military pension that you rated after twenty years.'[83] Correspondingly, just as Kennedy depended on deception and censorship in order to sustain the multifaceted appearances of personal vitality, public sanctity, academic excellence, literary skill, and military bravery upon which he had partly risen, so Glenn was forced to lecture fellow

---

81. van Riper, pp. 57, 202-11.
82. The Pierce and Schuon biography is a good example of this storyline.
83. van Riper, pp. 203-4; Wolfe, pp. 361-2.

astronauts and restrain overinquisitive reporters in order to protect the 'standard Apple Pie America' image the Mercury team had been given by *Life*. To this extent, Slothrop's fate at Neubabelsburg applied to the stars as much as it did to the starstruck: elevated by the seductions of imagery, both Kennedy and Glenn found themselves consumed inside the same system.[84]

Nor was this all. Glenn also discovered that he was expected to fulfil the more indefinite clauses in his contract for stardom: like the deities studied by Edgar Morin or like Charles Lindbergh, he had to live up to his own screen mythology, to satisfy the need for a type. John Kennedy may have enjoyed the attentions of his fans both for personal and electoral reasons. John Glenn and his fellow astronauts experienced the incessant demands of their own viewers as initially amusing and unnerving but ultimately irritating and exhausting expressions of something which moved beyond curiosity. Installed in their new, custom-built Houston homes, the Schirras, Grissoms, and Carpenters quickly found themselves besieged by tourists ripping grass from their gardens and fans begging photographs and signatures. Installed as the nation's premier space hero, Glenn himself soon began receiving letters from people seeking advice on investment strategies and the like. As he remarked in 1969, 'Somehow, once we'd flown, we weren't supposed to wash our cars or take out the garbage ... Suddenly we [were just] supposed to be experts on everything.' Forty years earlier Buster Keaton had signed a contract which prohibited him from laughing in public; in the early 1960s the astronauts signed no such document but nevertheless felt themselves equally tied up. They had become captives of their own fame and fortune.[85]

Yet such an outcome was only consistent with the necessities of an Operation which had created the Rocket State for its own ends and which would abandon it as necessary. No less than Blicero or Pointsman, the leading men of incipient totalitarian society were elevated in accordance with George Chicherin's

84. van Riper, pp. 148, 151-2; Garry Wills, *The Kennedy Imprisonment* (Boston: Little, Brown, 1982), pp. 16-34, 128-39; Herbert S. Parmet, *Jack: The Struggles of John F. Kennedy* (New York: Dial Press, 1980), pp. 323 ff; Herbert S. Parmet, *JFK: The Presidency of John F. Kennedy* (1983; New York: Penguin, 1984), pp. 17-18, 100-124; Victor Lasky, *JFK: The Man and the Myth* (1963; New York: Dell, 1977), pp. 476-80; Collier and Horowitz, pp. 153-5, 192, 202, 247, 254-5, 583.

85. Wolfe, pp. 358-9'; *New York Times*, 20 July 1969, p. 39; Morin, pp. 58-9, 66-8; Ward, pp. 5-6; van Riper, pp. 47-8.

earlier scheme: not as leaders but as agents of a power structure which would exploit and replace them once their usefulness, their holding pattern, was at an end.[86] In the case of John Kennedy the process of replacement was premature but, once the smoke had been cleared, could be seen to retain its own logic: elevated on a program of free enterprise and anti-communism, he was removed by operators *par excellence* for his failure to apply that program to Cuba and for his refusal to let others do likewise.[87] In the case of the astronauts their redundancy was equally clear. In an immediate sense, of course, they were produced as the ideal descendants of Whitney's interchangeable gun parts: one astronaut could and did step in for another where injury or accident required. But in the longer term, too, their contingent position was evident: popular at the Rocket State's apogee, they lost their appeal beyond. When in 1984, as the liberal mechanisms of the Rocket State's expansion continued to fragment under the pressures of their own contradictions, John Glenn attempted to reinsert his own established myth within the post-liberal political context (having himself described as 'Ike in a spacesuit' and as 'a potent mixture of John Wayne heroism, Buck Rogers fantasy, and Huck Finn Americana'), his presidential candidacy came to naught. Even the film *The Right Stuff* helped little. The dissolution of the Rocket State prompted public desires for vengeful policemen rather than intrepid pioneers. John Glenn had simply got the wrong dream.[88]

In 1961, when John Kennedy was presented with a vision of a manned moon landing as a fitting conclusion to his second term in office, the prospects had appeared fair; as it happened, the Rocket State failed to express itself fully in any such integral fashion. Events in Dallas in November 1963, and at Cape Kennedy in January 1967 (the Apollo 204 fire), destroyed both sides of the original equation, and in July 1969 it was Republican Richard Nixon, the political agent of the Rocket State's demise, who welcomed the first of the moon walkers home. To most Americans, however, the issue was secondary: the dead President's pledge had been fulfilled whilst the lives of the three martyred astronauts were recalled as the tragically necessary

86. See above, p. 61.

87. I rely here on Anthony Summers, *Conspiracy* (London: Victor Gollancz; Glasgow: Collins/Fontana, 1980).

88. Mailer, *Fire on the Moon*, p. 40; van Riper, pp. 22-3, 25.

costs of progress. Through all events the Operation persisted, with *Life* not only providing illustrations and text but also illuminating the essential relationships between the agents and procedures of power. During the earliest days of the Rocket State, in 1940, its founder had provided Kennedy with a start to his public career by writing an introduction to *Why England Slept*; during the 1950s, as the American Century matured, it had promoted the Senator both as celebrity and as politician; in 1960, as Luce's program flourished, it had effectively helped bring him to high office. Three years later *Life*, unlike its ostensible beneficiary, remained fully operative: not simply prepared to lament his passing but financed to outbid all others for the best color film of the President's murder in preparation for its memorial issue.[89]

---

89. Sorensen, p. 525. According to a former *Life* employee, Time Inc. paid Abraham Zapruder $150,000 for 'exclusive world rights to the photographs, both as stills and as movie footage. The decision and the price both came about from a genuine desire on the part of Zapruder and *Life* to tie up the precious film tightly so that it could not fall into the hands of unscrupulous profiteers [sic].' Dora Jane Hamblin, *That was the Life* (New York: Norton, 1977), p. 176. As one researcher into Kennedy's murder reports, *Life* not only enjoyed vast sales returns from its editions carrying parts of Zapruder's footage but also doctored the presentation of the prints in order to support some of the leaked conclusions of the soon to be published Warren Report into the assassination. See Robert Sam Anson, *They've Killed the President* (New York: Bantam, 1975), pp. 83, 131-4, 138-9.

# 5

# Behind the Screens

## Happyville: A Dream of the Future's Face

As we begin to master the potentials of modern science we move toward a new era in which science can fulfil its creative promise and help bring into existence the happiest society the world has ever known.

— John F. Kennedy (22 October 1963)

What we will have attained when Neil Armstrong steps down upon the moon is a completely new step in the evolution of man. It will cause a new element to sweep across the face of this good earth and to invade the thoughts of all men.

— Wernher von Braun (15 July 1969)

At jet-set parties in Paris, grand tribal fires of Southern Zambia, in the courtyards of Buddhist temples in Bangkok, on street corners in Colombo, Ceylon, and in snug Dublin pubs millions huddled close to TV sets or radios as the Apollo voyage was described in dozens of languages. For days newspapers, radios, and television stations throughout the world have featured the historic space journey, often giving little attention to domestic issues. It seemed as if there were no part of earth unaware that two men had set foot where no man had gone before.

— *New York Times* (21 July 1969)

From the pictures of John Kennedy and the Mercury astronauts in the White House Rose Garden in 1961, to those of understudy Richard Nixon and the Apollo XI crew on board the recovery ship *USS Hornet* in 1969, the victorious, smiling faces of the Rocket State's leading men elicited an appropriate audience response. Of course the Atlas workmen who 'cheered their brains

out' for Virgil 'Gus' Grissom when he addressed them at
Convair's San Diego plant, like the welding inspector who took
great pride in being introduced to John Glenn at NASA's Langley
Research Center in Virginia, had immediate material grounds for
their enthusiasm. That they were amongst the first of the space
program's many satisfied customers was perhaps not surprising.
And yet their 'beaming mug[s]' were quickly reproduced across
the face of the entire nation: in the expressions of the hundreds of
servicemen who smiled and stared at Alan Shepard as he was
plucked from his capsule off Bermuda; in the reactions of the 500
Congressmen who wept and cheered for John Glenn as he spoke
to an emotional House of Representatives; in the faces of the
millions who stood 'crying and waving little flags and pouring
their hearts out' as the limousines parading the country's astronauts inched their richly upholstered way to New York City
Hall.[1]

Moreover, while the space agency's labor force amounted by
1969 to little more than half of its peak size and its federal budget
had been cut by a quarter, such reductions had negligible impact
on the extension of the Rocket's appeal. On the contrary, as
Apollo XI was prepared for launch unprecedented numbers took
their seats for its eight-day-long epic. At Cape Kennedy Norman
Mailer witnessed diversity made one, as nameless southern mill
workers and west coast aerospace executives, New York news-
papermen and 'families of poor Okies' drew up along the beach
while union leaders, princes, diplomats, soldiers, cardinals, and
comedians formed a 'mafia of celebrity' in the stands. Further
afield the *New York Times* reported millions more tuning in as the
mission progressed: 'From Wollongong, Australia, where a local
judge brought in a television set to watch while hearing cases, to
Norwegian Lapland, where shepherds tended their reindeer with
transistor radios pressed to their ears.' A few days before at the
Cape several hundred journalists and photographers had been
left 'standing on tiptoe at every bad angle and hopeless vantage-
point' for a 'chance of the briefest glimpse' of the crew; now
several hundred million people went out of their way to see their

---

1. Tom Wolfe, *The Right Stuff* (London: Jonathan Cape, 1979), pp. 148, 266-
7, 345-52; Loyd S. Swenson, Jr., James M. Grimwood, and Charles C. Alexander,
*This New Ocean: A History of Project Mercury*, NASA SP-4201 (Washington, D.C.:
National Aeronautics and Space Administration, 1966), p. 241; Philip N. Pierce and
Karl Schuon, *John H. Glenn: Astronaut* (New York: Ballantine Books, 1962), pp. 131-6.

lunar excursion. In London Queen Elizabeth stayed up through the night; at Castel Gandolfo Pope Paul VI surveyed the moon through his telescope; in Washington the President remained 'glued to his set' for hours. The March 1961 forecast of the National Academy of Science's Space Advisory Board – that 'man's exploration of the moon [was] potentially the greatest inspirational venture of the century and one in which the whole world can share' – seemed vindicated at last. As the *New York Times* front page headline put it on 21 July, 'ALL THE WORLD' was 'IN THE MOON'S GRIP.'[2]

Such deep-seated, extensive, and voluntary integration appeared both to satisfy and vindicate the demands of the occasion's original advocate. Radio Free Europe's agitation and the tactics of militarized rollback in the Far East had previously yielded little more than one performance of the Bob Hope show in Pyongyang and equally ephemeral hope in Budapest. Kennedy's universalist showpiece now captured imaginations across the 'Free World,' the third world, and the eastern bloc alike. As one Czech broadcaster put it: 'One Frank Borman on our television screens outweighs 100 communist bigwigs bussing one another on the cheeks and laying wreathes.'[3] Notwithstanding the fact that it was Eisenhower's former Vice-President rather than Kennedy himself who publicly heralded this integration, when Nixon proclaimed in his July 1969 address to the planets that 'for one priceless moment in the whole history of man, all the people on this earth are truly one,' he articulated the essential objective of the incipient totalitarian dream last advocated as the Kennedy promise: the reconciliation, not merely of former electoral adversaries like Lyndon Johnson and Barry Goldwater (whose pe-launch handshakes Norman Mailer recorded in *Of A*

---

2. Thomas P. Murphy, *Science, Geopolitics, and Federal Spending* (Lexington, Mass.: D.C. Heath, 1971), p. 364; Arnold S. Levine, *Managing NASA in the Apollo Era*, NASA SP-4102 (Washington, D.C.: National Aeronautics and Space Administration, 1982), p. 107; Norman Mailer, *Of A Fire on the Moon* (Boston: 1970; New York: New American Library, 1971), pp. 59-61, 80, 84-5; *New York Times*, 16 July 1969, p. 22, 21 July 1969, pp. 1-2, 22 July 1969, p. 1; Walter A. McDougall, *The Heavens and the Earth* (New York: Basic Books, 1985), p. 315.

3. Callum MacDonald, *Korea: The War Before Vietnam* (London: Macmillan, 1986), p. 210; *New York Times*, 16 July 1969, p. 22; Blanche Wiesen Cook, 'First Comes the Lie: C.D. Jackson and Political Warfare,' *Radical History Review* 31 (1984), pp. 50-3, 63-4. Frank Borman was the Commander of Apollo VIII, the first manned circum-lunar mission of December 1968.

*Fire on the Moon*), but of much deeper 'violent political opposites' whose progressive alignment now tightened the Rocket State's grip.[4] In fostering the total commitment of private energies to what was the most absorbing of all public works, the production and consumption of Apollo XI endeavored to extend the Rocket State into the farthest reaches of Pynchon's Zone: inaugurating a terminal society devoid of contradictions whose boundaries were no longer local – as they had been when Tom Wolfe saw John Glenn's victory parade converting the 'horrible rat-grey city' of New York into a community both touching and warm – but global; a universal Happyville where all hearts and minds found features in common beneath the Orpheus Theater's panoplied chambers (*GR* 318).[5]

That dream was explicitly articulated by James Earl Webb, NASA's Administrator for over seven years following his appointment by Kennedy in February 1961. To Webb the large-scale integration of men and machinery, of invention and production carried out by the space agency in the execution of the Apollo project comprised both a practical mechanism for the creation of an efficient, benevolent, and liberating social system and a formal paradigm for the rational administration of the resulting society. By combining the engines of private enterprise with the guidance systems of the state, the consumables of industry and the monitors of academia with the impulse powers of science and technology, modern executive systems such as those pioneered by NASA could, 'on an almost fixed time schedule, meet new needs or effect desired improvements in our situation as a people and as a nation.' Whether it was in the realm of crime prevention or urban transportation, Webb wrote in his *Space Age Management*, Americans had 'the ability ... to accomplish almost any task' they might set for themselves. Since there was 'no reason why we cannot do in other areas what we have done in aeronautics and space,' there was every reason to believe that 'the NASA system ... might become the pattern needed by the nation' in the years to come. In sum, as Walter McDougall concluded from his study of Webb's published and private remarks, NASA's long-serving

---

4. *New York Times*, 21 July 1969, pp. 1-2; Mailer, *Fire on the Moon*, p. 84. Cf. Hannah Arendt, *The Origins of Totalitarianism*, new edn (New York: Harcourt Brace Jovanovich, 1973), pp. 438, 465-6.

5. Tom Wolfe, p. 349.

Administrator believed that the American space program could furnish the mechanisms for 'revolution from above.'[6]

Such a 'revolution' would be driven forward, and would have its wholesome character guaranteed, not by the outdated and laborious processes of class struggle but by the rejuvenating powers of the eternal frontier. For whereas doctrines of social conflict bred only tyranny, experience of life at the outposts fostered both democracy and harmony in the face of common hardships. Claiming support from the theories of historians Frederick Jackson Turner and Walter Prescott Webb, NASA's chief executive saw the new, uncharted reaches of space as vast extensions of the old, unsettled territories of the west. Consequently, by placing 'an entire nation ... as an organized entity in the same position as the pioneer was individually on the frontier,' the space agency's operations would necessarily promote peace and cooperation, not only at home but by stages, as other nations joined the United States on that final frontier, around the world.[7] Indeed, they would do more. As Webb's successor at NASA, Thomas O. Paine, implied a few days before the establishment of Tranquillity Base, the new frontier's justice had no need to cease at the planet's edge. On the contrary, 'continuing advances in space transportation offer[ed] man new alternatives for free societies' stretching out to the stars and beyond:

> As with the American experience of 1776, founding a new society in a demanding environment will sweep aside old world dogmas, prejudices, outworn traditions, and oppressive ideologies. A modern frontier brotherhood will develop as the new society works together to tame its underdeveloped planet for posterity ... Orwell's *Nineteen Eighty-four* may still haunt us, but ... the specter of a monolithic imperium over all mankind may well be receding by 1984 if America has the vision and determination to press forward in Apollo XI's fiery wake.[8]

The argument was familiar enough, variations on its theme having been advanced during the 1950s and 60s by a number of interested parties plying the space between science's fiction and

---

6. James E. Webb, *Space Age Management. The Large-scale Approach* (New York: McGraw-Hill, 1969), pp. 21-2, 24; McDougall, pp. 381-6.

7. McDougall, pp. 387-8.

8. *New York Times*, 17 July 1969, p. 47.

facts.[9] Rendered particularly visible in the form of NBC Television's *Star Trek*, whose American-led multinational corps of
missionaries began spreading peace by enterprise in September
1966, it would remain familiar for as long as its mythologically
sanctioned validity could be drawn on, as it was by Paine in 1969,
to justify the exploitation of post-lunar space.[10]

If the relationship between the large-scale organizations which
produced and maintained James Webb's well-ordered society
and the frontier spirit which supposedly inspired it appeared
obscure, then neither Webb nor Paine troubled themselves to
investigate it further in public. Both men operated within the
understanding that, formal postures notwithstanding, the attractions of the space program and of the Rocket State in general lay,
as they did with the astronauts, in their evident reliability and
practical application rather than in their abstract constitution.
Webb's digressions on the potentials of space age management
might have reactivated Franz Pökler's earlier interest in the possibilities of a 'Corporate City-State' or engaged Clive Mossmoon's
administrative eye; likewise, Paine's exercise in historical analogy
could have provided Gerhardt von Göll with the germ of an idea.
But there were other figures, named and nameless, who survived
the ruins of the Oven State in *Gravity's Rainbow* and, as top NASA
officials well knew, there were other interests for the Rocket State
to enlist.

For many, enlistment proved unnecessary. Clayton Chiclitz's
early impressions of the rocket's potential – 'I think there's a
great future in these V–weapons. They're gonna be really big' –
may have taken time to be appreciated fully by his contemporaries but, with the approval of NSC–68 and the von Neumann
and TCP Panel reports between 1950 and 1955, his foresight was
progressively realized. By 1957 the launch of the Sputniks would
stimulate more than just toy manufacturers and telescope sales.
Within months the President of the McDonnell Aircraft Corporation was enthusiastically welcoming his 'fellow pilgrims' to 'the
wondrous age of astronautics.' Thereafter, as the new National

---

9. See, for example, Arthur C. Clarke, *Interplanetary Flight: An Introduction to
Astronautics* (London: Temple Press, 1950), pp. 144-6; H.K. Kaiser, *Rockets and
Space Flight*, trans. Alex Helm (London: Faber & Faber, 1961), pp. 141-2; Frank B.
Gibney and George T. Feldman, *The Reluctant Space-Farers* (New York: New Ameri-
can Library, 1966), pp. 151-72.

10. See Stephen Whitfield, *The Making of Star Trek* (New York: Ballantine, 1968),
pp. 22-6, 202-5.

Aeronautics and Space Administration budget rose from $223 million in Fiscal Year 1959 to $1830 million in FY 1962, during which time McDonnell secured prime contracts for both Mercury and Gemini spacecraft, the captains of industry volunteered their services in ever greater numbers.[11] Between September 1961 and November 1962, North American Aviation was chosen as prime contractor for the Apollo Command and Service Modules, the second stage (S–II) of the Saturn V rocket, and both the latter's F–1 and J–2 engines; Boeing secured an order to build the first stage of the launching vehicle (S–IC) whilst Douglas Aircraft made sure of the third (S–IVB); Grumman won the contest to construct a Lunar Excursion Module; and General Electric were commissioned to provide integration analysis, reliability testing, and checkout systems for the entire project. Clayton Chiclitz' hunch proved both literally and metaphorically correct. The Mercury program eventually cost barely $400 million in all, but these seven main awards would by 1969 be worth nearly $10 billion alone. In little more than a year, therefore, Webb and his colleagues had secured the commanding heights of the aerospace industry. The Rocket State's engines were primed (*GR* 558).[12]

Where McDonnell, North American, and other prime contractors led, hundreds of smaller firms making thousands of lesser deals followed in subcontractual train. Between 1958 and 1963 Project Mercury provided work for 12 prime contractors, 75 first order subcontractors, over 1,500 second tier and around 7,200 third tier suppliers; between 1961 and 1973 the Apollo program drew on the resources of at least 20,000 and possibly as many as 80,000 firms at all levels. The sheer extent of such subcontracting, financed by a space agency budget which rose to over $5 billion per annum during the mid-1960s, in turn ensured the accommodation of not only Clayton Chiclitz' less corpulent corporate colleagues but also the bulk of the Rocket State's crew. Even though four industrial giants (North American, McDonnell, Grumman, and Boeing) secured over half of the agency's

11. James McDonnell, 'The Conquest of Space: A Creative Substitute for War,' speech, Washington University, St. Louis, June 9 1958, quoted in Swenson, *et al.*, p. 119; Murphy, p. 364.

12. Levine, pp. 319-20; Richard Thruelsen, *The Grumman Story* (New York: Praeger Publishers, 1976), pp. 331-3; Mary A. Holman, *The Political Economy of the Space Program* (Palo Alto, Cal.: Pacific Books, 1974), pp. 69, 74; Swenson, *et al.*, pp. 508, 643; Murphy, p. 173.

cumulative dollar outlays up to June 1966, it was the subcontractors – components manufacturers, overflow firms, service
agencies and the like – rather than the giants themselves who
provided nearly three-quarters of the agency's workforce, from
Pökler's fellow experts and Mossmoon's massed management to
the rank and file spin-offs from Tyrone' Slothrop's decay. As the
elaborate edifice called into being by the decision to go to the
moon gained weight, therefore, the main engines on which it
would rise not only extended their lines of supply to its more
peripheral means of production but also forged a pan-classical
personnel alliance: blending mechanics and managers, scientists
and secretaries, directors, designers, and more to weld the Rocket
State's main stages together.[13]

This process of comprehensively beneficial integration was
perhaps most clearly displayed in the south, where a combination of factors – ranging from mild climate, established facilities,
and industrial, personnel, and energy reserves to the potential for
seaborne and airborne access, educational and social provision,
and relative safety – led NASA to locate almost all of its main
installations: the Marshall Space Flight Center in Huntsville,
Alabama, the Michoud Assembly Facility near New Orleans,
Louisiana, and the Mississippi Test Facility at nearby Pearl
River, which together turned blueprints into functioning
boosters; the Launch Operations Center (later Kennedy Space
Center) at Cape Canaveral, Florida, which assembled and fired
the various spacecraft; and the Manned Spacecraft Center near
Houston, Texas, from which missions were tracked and
controlled.[14] As more than one observer noted, the location of
these and other smaller facilities gave the American space
program a distinctly southern accent: personally, through the
prominence of men like Administrator Webb and chief NASA
spokesman Julian Scheer, both of whom hailed from North
Carolina; materially, through the absorption by the four Gulf
states hosting the larger space centers (Florida, Louisiana,

13. Holman, pp. 68, 77, 84, 281-2; Richard J. Barber, *The Politics of Research*
(Washington: Public Affairs Press, 1966), p. 79; Murphy, p. 364; Swenson, *et al.*,
pp. 17, 508; Gibney and Feldman, p. 113.
14. Redstone Arsenal was redesignated the George C. Marshall Space Flight
Center on 1 July 1960. The other four installations were announced by NASA
between 24 August and 25 October 1961. The selection procedures for these
facilities are detailed and evaluated in Holman, pp. 43-62; Murphy, pp. 197-223.

Alabama, and Texas) of nearly a quarter of all NASA expenditure – over $6 billion in all – between Fiscal Years 1962 and 1969. Yet, as Webb and his colleagues also knew, this regional emphasis was at the same time intended to give southern society as a whole a more conventionally American tone by extending the Rocket State to pastures new.[15]

Extensive space-related investment in the south would, NASA advocates hoped, play a catalytic role in fostering the progressive modernization of a region which still harbored the most backward of American societies. On the one hand such expenditure was designed to appease the unreconstructed south by providing its beneficiaries and defenders with the prospects of larger agricultural markets, inflated land values, and a general sense of identification with national achievement in return for its toleration of Washington's efforts to extend limited political concessions to the most glaring of Michael Harrington's 'other Americas.' Desegregation, that is, would be balanced by the spoils of prosperity and purpose. On the other hand, investment in space was intended to undermine the structural significance of that same old order by nourishing the growth of a new, urban, and industrialized south no longer dependent on the peculiar institutions of its predecessor.[16] Where the reactivation of the Banana River Naval Air Station (later Patrick Air Force Base) in 1947 and of the US Army Ordnance Corps' Redstone Arsenal in 1949 had helped boost the annual average rate of employment growth in Brevard County, Florida, and Huntsville, Alabama, by 400 per cent and 100 per cent respectively between the 1940s and 50s, so by initiating what *Fortune* described as 'Dixie's greatest job rush,' the expansion of federal investment through the space agency would foster the growth of a southern technocratic and university oriented middle class divorced from the agrarian elite. If moribund mill towns like Huntsville could be reconstituted as

15. Loyd S. Swenson, Jr., 'The Fertile Crescent: The South's Role in the National Space Program,' *Southwestern Historical Quarterly* 71 (1967/68), pp. 377, 390; statistics derived from Table 6.14 in Murphy, pp. 183-5.

16. Leonard Reissman, 'Urbanization in the South,' in John C. McKinney and Edgar Thompson (eds), *The South in Continuity and Change* (Durham, N. Cal.: Duke University Press, 1965), pp. 84-5; McDougall, p. 376. See also President Kennedy's 12 September 1962 speech, 'Science, Space and the New Education,' given at Rice University, Houston, in Allan Nevins (ed.), *The Burden and the Glory* (New York: Harper & Row, 1964), pp. 241-6.

the dynamic poles of a modern and high growth economy, then the reaction engines they designed might, in Loyd Swenson's words, 'help the south to rise again and to leave behind reactionary thought.' As James Webb put it in the autumn of 1963, with the assistance of the space agency anachronisms of the past would steadily acquire 'profiles of the future.[17]

Evidence provisionally substantiating Webb's optimistic projections accumulated over time across the full width of NASA's 'fertile crescent.' Thus, as the giants of the aerospace industry and their subcontractual posses started to follow the federal bandwagon south during the early 1960s so the *New York Times* reported their combined relocation turning 'great spaces of wasteland into humming industrial areas.' In Huntsville the Marshall Space Flight Center created 18,600 new jobs directly and another 20,000 indirectly between 1960 and 1966, contributing significantly to an average annual employment growth rate of 11.5 per cent over the same period. At Cape Canaveral the $700-million extension of the Atlantic Missile Range carried out between 1961 and 1966 created 20,000 new jobs directly and well over 30,000 indirectly, which together accounted for some 56.3 per cent of Brevard County's total employment during the summer of 1966. In Hancock County, Mississippi, the establishment of the booster test facility quickly reduced an unemployment rate of 15 per cent following years of stagnation by creating 9,000 new jobs – mostly with General Electric – between 1963 and 1965. In New Orleans the reopening of the Michoud Assembly Facility created upwards of 11,000 new jobs with Boeing and Chrysler, reversed a five-year employment slump, and made the Apollo program the largest single industrial employer in Louisiana by 1966, accounting for 8 per cent of the state's total manufacturing earnings and an $80 million annual payroll. Even in Houston, where the oil, gas, and petrochemicals industries that ensured local prosperity made the new NASA facility more an object of civic pride than of individual economic significance, the coming of the Manned Spacecraft Center brought with it over

---

17. Paul O'Neill, 'The Splendid Anachronism of Huntsville,' *Fortune*, June 1962, pp. 151 ff.; Holman, pp. 202-3; Ben B. Seligman, *The Potentates: Business and Businessmen in American History* (New York: Dial Press, 1971), p. 362; Swenson, pp. 382, 391; James E. Webb, 'Profiles of the Future: The Economic Impact of the Space Program,' *Business Horizons* 6 (1963), pp. 4-26.

100 aerospace firms both large and small within a year of its ground-breaking ceremonies, and nearly 13,000 direct new jobs thereafter.[18]

These developments were not merely quantitative, however. If the space agency's southern location decisions indirectly helped discharge its statutory duty to channel federal contracts towards regions of persistent 'labor surplus,' then they also helped implement its less formal social purpose by extending the circuits of the domestic high wage economy beyond the unionized walls of the industrial heartland. The visible future which James Webb profiled did not, after all, include masses of the laboring poor amongst its prominent or permanent features. Rather, it described what NASA salaries delivered: the inception of middle-class ease. In New Orleans employees at the Michoud Facility earned on average $8,000 a year in 1966 compared to the city's mean manufacturing wage of $6,800 per annum. In Houston non-technical civil service personnel at the Manned Spacecraft Center earned $10,000 per annum on average during 1966 whilst typical non-supervisory workers in non-NASA manufacturing jobs made $6,700 a year. Comparable income ratios obtained around NASA's three other large southern facilities. In turn, the reproduction of such incomes (recorded in both Brevard County and Houston between 1959 and 1966 by rates of growth in total personal income of over double the national average) helped further a more general tendential homogenization of the New South's axes and the rest of the nation as a whole. As per capita income levels in places like Huntsville moved towards the national average during the 1960s; as changes in the class distribution of income both around NASA installations in particular and across the south in general reduced income inequalities and bolstered the absolute and relative size of the middle classes; and as various indices of southern development – from the improvement in educational facilities and the boom in general construction work to the expansion of retail sales – moved upwards, so the domain of the Rocket State and its elevator network spread. Whether it was in Huntsville, where in

18. *New York Times*, 29 October 1963, p. 34; Swenson, pp. 382, 388-9; Holman, pp. 200-203, 215, 225-8, 355; Murphy, p. 216; Stephen B. Oates, 'NASA's Manned Spacecraft Center at Houston, Texas,' *Southwestern Historical Quarterly* 67 (1963/64), p. 372.

the mid-1960s residents began vigorously protesting against Alabama segregation practices, or in the Manned Spacecraft Center near Houston, which in 1969 Norman Mailer found awash with descendants of some union between Mossmoon and Pökler, Webb's incipient totalitarian order progressively made its face public.[19]

This integrated structure of production, exchange, and consumption – of burgeoning industrial activity, expanding domestic liquidity, and rising consumer demand – had its limits, of course. Technological, political, economic, and strategic factors aside, the requirements of the Buy American Act alone restricted any interest NASA might have had in directing work overseas to small contracts for intimate allies – such as the deal which furnished the first men on the moon with British-made water-cooled underwear. But if the Rocket State's structure of production remained bound by a relatively privileged circle, its structure of consumption nonetheless found room to install all comers. As the *New York Times* reported on the eve of Apollo XI's lift-off, Europeans were 'as excited as many Americans and ... feel as personally involved.' The French, usually 'so tired of politics and world affairs,' were addicted; the Dutch 'lunar-crazy.' Neil Armstrong's first small step saw even the iron curtain go up. In Warsaw the US Embassy was packed with 1,000 cheering Poles; in Cracow a statue was unveiled to the lunar pioneers; in Prague streets were deserted as crowds of Czechs clustered round televisions in beer gardens and hotel lobbies; in East Berlin the Apollo VIII capsule was displayed to the admiration of Soviet troops. Across the planet comparable scenes were acted out; from Cuba, where the jamming of Voice of America transmissions was lifted, to Djakarta, where unhindered US Information Service broadcasts led to Americans being stopped in the streets and congratulated; from Australia, where trading was halted on the Sydney Stock Exchange, to Italy, where crime rates in Milan

---

19. Holman, pp. 77-8, 206-8, 211-14, 221-9; Oates, 370-75; Gibney and Feldman, pp. 116-17; Mailer, *Fire on the Moon*, pp. 15-16. The developments described here were not, it need hardly be emphasized, pervasive, either regionally or structurally. Even as the combined efforts of CORE, SNCC, and SCLC to eradicate segregation in public places and to extend voter registration in the south made progress, so the old order held its ground. While more schools were built and more teachers employed, local political elites moved existing institutions towards desegregation with all the deliberate speed of a Saturn V transporter.

went down by two-thirds.[20]

In this sense, everyone had something in common. Good television, even the flickering images of Armstrong and Aldrin's futuristic waltz, was the inalienable right of all mankind. Moreover, if the Orpheus Theater's elevator network remained off-limits to those wasting away in its 'dark, unheated [and] somehow forbidden levels,' then the 'common aerodynamic effect' of its 'intimate cubic environment' – the 'Promise of Space Travel' first glimpsed in the Oven State's slave zones – alerted everyone to its modernized trim. For the theater's first circle of customers the promise was of technological novelties to buttress the 'pin-ups and library shelves' of their ever more personalized vantage-points: in the words of NASA Deputy Administrator Hugh L. Dryden, 'a great variety of new consumer goods and industrial processes that will raise our standard of living and return tremendous benefits to us in practically every profession or activity.' For the less well-situated the promise was, as James Webb argued, of space age technology turned to community use: in fire fighting and drug traffic control, in urban redevelopment and the care of the mentally ill. And for 'all those good folk in the queue,' as night manager Richard M. Zhlubb described the denizens of the cold and dark, the promise was of guides to good seats in the future: weather satellites and resource satellites, educational satellites and communications satellites, all of which would, in President Kennedy's words, foster 'knowledge and understanding' to be 'used for the progress of all people' (*GR* 296-7, 450, 735-6, 754-5).[21]

---

20. Levine, p. 99; *New York Times*, 16 July 1969, p. 22, 17 July 1969, p. 26, 21 July 1969, p. 10, 22 July 1969, p. 1; *Sunday Mirror* (London), 20 July 1969, p. 17. Communist China was the most significant and populous exception. There, the *New York Times* noted with smug neutrality, as Armstrong stepped out onto the lunar surface 'Peking Radio was broadcasting an account of how workers in a Shanghai factory had benefitted from the teachings of Mao Tse-Tung.' That the United States on its part did not officially recognize 800 million Chinese at all was left unrecorded.

21. McDougall, pp. 383, 385, 421; Nevins (ed.), p. 243. See also President Kennedy's 24 July 1961 statement on US satellite policy, *Public Papers of the President, 1961* (Washington: US Government Printing Office, 1962), p. 530. Two projections of global society under the influence of satellite technology – one cautiously optimistic, the other unreservedly so – are Lester R. Brown, *World Without Borders* (New York: Vintage, 1973), pp. 259-77; and Arthur C. Clarke, 'Everybody in Instant Touch,' in Arthur C. Clarke (ed.), *The Coming of the Space Age* (1967; London: Panther, 1970), pp. 150-63.

Paraded up and down at intervals for over a generation, a tempting selection of sidelines and spin-offs gave such promises material expression. Advances in everything from microelectronics to thermal underwear, from freeze-dried food to cordless razors, and from scientific monitoring equipment to cigar-sorting machinery satisfied not only the consuming ambitions of von Göll's, Greta Erdmann's and Tyrone Slothrop's more energetic offspring but also the jaded appetites of such retiring past masters as Pointsman and (the late) Brigadier Pudding. Studies carried out between 1964 and 1965 by Aerojet–General and North American Aviation into the development of smog control systems for Los Angeles and state-wide public transport for California defined projects tailor-made for the application of fuel cell technology introduced by the flights to the moon. And the diffusion during the 1970s and 80s of Landsat data in particular – with ground stations and applications teams being set up across Africa, Asia, and Latin America – made possible a form of resource control unthinkable in the pre-Sputnik era: whether in Thailand, where Landsat pictures provided information for forestry and staple crop administration, or in Guatemala, where they helped locate an extensive new basin of oil.[22]

In promising to deliver life to barren lands and liberty to captive peoples, and to facilitate the pursuit of Happyville for all, the American space program beckoned closer the day when all mankind would become links in the machinery of freedom. To the businessmen of America, operating under the responsible yet unobtrusive eye of the space agency and the criteria laid down by its congressional guardians, what the enterprising hero of *Destination Moon* described as 'the greatest challenge ever hurled at American industry' brought not merely business but the opportunity to demonstrate how private enterprise could simultaneously respond to the demands for mobilization engendered by Sputnik, serve the public interest by delivering long-term economic growth and higher employment, and vindicate the claims of the Free World's defenders. To those hundreds of thousands of men and women involved directly in the manned

22. Holman, pp. 326-31; Gibney and Feldman, pp. 125-7; Barber, pp. 82, 149; Christopher Rand, *Los Angeles* (New York: Oxford University Press, 1967), p. 97; John McHale, 'Big Business Enlists for the War on Poverty,' *Trans-action* 2, no. 4 (1965), pp. 3-4; Martin Ince, *Space* (London: Sphere, 1981), pp. 111, 120.

space program, from the beaming welding inspector at Langley to the still grinning Grumman workers building the last of the Lunar Modules at Bethpage, it brought not only jobs and incomes, skills and pride, but also the security of work under federal law: minimum wage and maximum hours regulations, equal opportunity provisions, and union rights. To the uncounted millions indirectly tied in, from the Los Angeles ice-cream manufacturer who served up a line in 'lunar cheesecakes' and the Las Vegas nightclub stripper who worked down from a simulated spacesuit, to their hungry customers, it brought either profitable inspiration or expensive indulgence, the fruits of their freedom to choose. To the biggest television audience in history (between one-fifth and one-third of the world's 3.27 billion people in July 1969), it brought a mixture of exhilaration and exhaustion, which made not only the viewers but also the engineers and entrepreneurs, bureaucrats and beaming employees whose work was already done at one with the astronauts themselves.[23]

Sheltered by an elaborate construction built into the very foundations of the Rocket State, such freedoms were defined and elevated in turn as the manifest lineaments of nature. Where in 1945 the production of Clive Mossmoon had inaugurated the reproduction of an increasingly dehistoricized post-war society, now that society's crowning lunar achievement was widely celebrated as the faithful execution of some pre-ordained natural order. To the *New York Times* the lunar landing constituted 'more than a step in history; it [was] a step in evolution.' To Buckminster Fuller it marked 'the dead center of evolutionary events.' To Wernher von Braun it ranked 'equal in importance to that moment in evolution when aquatic life came crawling up on the land.' Correspondingly, the agents of such an imperative action were redefined as molds of sociobiological necessity, with Armstrong becoming merely the latest voluntary expression of what the *New York Times* described as an 'urge ... deeply

---

23. Murphy, pp. 45-6; Mailer, *Fire on the Moon*, p. 138; Irving Pichel, dir., George Pal, prod., *Destination Moon*, script by Robert A. Heinlein, Eagle-Lion, 1950; Leonard Silk, 'The Impact on the American Economy,' in American Assembly, Columbia University, *Outer Space: Prospects for Man and Society* (Englewood Cliffs, N.J.: Prentice Hall, 1962), pp. 76-7; Wolfe, p. 148; Thruelsen, pp. 358-61; Holman, pp. 77, 80, 163, 191, 270-71; Levine, pp. 94, 99-100, 107, 120-21; *New York Times*, 18 July 1969, p. 67, 22 July 1969, pp. 1, 27.

inscribed in man's psyche' that drove him on as it had driven Lindbergh and Columbus in the past. As Armstrong himself remarked shortly before his flight, if the voyage to the moon had been made possible only 'because of the success of four previous Apollo command flights,' then each astronaut was ultimately driven skyward 'by the nature of his deep inner soul.' They were 'required to do these things just as salmon swim upstream.'[24]

Such equations of liberty and natural order were made from a position of absolute faith. As Hannah Arendt pointed out in 1963, seen from the ultimate vantage-point of space – that point of Faustian *Übersichlichkeit* which General Cummings imagined in *The Naked and the Dead* in 1948 and which Lyndon Johnson pushed the US towards from 1958 onwards – the contingencies of history would indeed take on the appearance of a 'large-scale biological process,' while the freedom which Sartre defined in the late 1950s as 'the irreducibility of the cultural order to the natural order' would equally be seen as 'no more than overt behavior' rather than 'the result of conscious human effort.' Were an entire society so highly advantaged, it followed, its characteristic liberty might also be taken as the exercise of instinctive necessity and any 'revolution from above' accepted as no more than a sign of evolution down below. This design, encapsulated in *Gravity's Rainbow* by Pynchon's image of the rocket's 'scheduled parabola' becoming as natural as the rainbow's curve, marked the culmination of a long engineering tradition built up and progressively revised inside the Operation's leading circles by Walter Rathenau and George Chicherin, Gerard Swope and Franklin D. Roosevelt, B.F. Skinner and James E. Webb amongst others, and perhaps topped out by Zbigniew Brzezinsky, the Rockefeller confidante, Trilateral commissioner, and presidential adviser who, even as the moon landings commenced, perceived mankind to be en route towards a society 'that [was] shaped culturally, psychologically, socially, and economically by the impact of technology and electronics – particularly in the area of computers and communications.' It was the vision of a terminal society engineered by a patterning elite in which all private and public life would be absorbed and governed as civil and sanctioned behavior; a vision

---

24. *New York Times*, 20 July 1969, sec. 4, p. 12, 21 July 1969, pp. 6, 11; Mailer, *Fire on the Moon*, pp. 26, 43-4, 69.

which for Pynchon became a preview of the Rocket State in full occupation and for Mailer 'a dream of the future's face.'[25]

## Pain City: The Amputated Limb

> Kekulé dreams the Great Serpent holding its own tail in its mouth, the dreaming serpent which surrounds the world. But the meanness, the cynicism with which this dream is to be used.

> — Thomas Pynchon, *Gravity's Rainbow* (1973)

> A certain kind of business has to be done behind the door.

> — Bobby Baker, *Wheeling and Dealing* (1978)

As Pynchon and Mailer both understood, that dream reposed on a nightmare of its own dissolution. Notwithstanding the elevated prose of John Kennedy's inaugural address – 'together let us explore the stars' – the spectacle of a national, bilateral, and ultimately global space age unity articulated by the President and his colleagues during the 1960s remained predicated on limitless conflict. Between 1957 and 1959, it was Soviet space success, not *Explorer 1*, which gave rise to NASA and the manned space program; in 1961 it was the achievement of Gagarin, not of Alan Shepard, which primarily spurred the drive to the moon. Correspondingly, it was Khrushchev's denial of Soviet moon landing plans which prompted congressional moves to reduce NASA's FY 1964 budget request by 10 per cent, and Kennedy's renewed proposal for a joint lunar mission which led to a ban on any such mutual projects in 1963. Cooperative ventures undertaken within these terms may have received extensive official promotion, but they were strictly limited to initiatives in space science, weather data exchange, and the like, and did not include

25. Hannah Arendt, 'The Conquest of Space and the Stature of Man,' in *Between Past and Future*, enlarged edn (1968; New York: Penguin, 1977), p. 279 (first published in *American Scholar* 32 (1963), pp. 527-40); Lyndon Johnson, speech before Senate Democratic caucus, quoted in Doris Kearns, *Lyndon Johnson and the American Dream* (London: André Deutsch, 1976), p. 145; Jean-Paul Sartre, *Search For A Method*, trans. Hazel E. Barnes (1963; New York: Random House, 1968), p. 152; Zbigniew Brzezinsky, *Between Two Ages: America's Role in the Technetronic Era* (1970; New York: Penguin, 1976), p. 9; Mailer, *Fire on the Moon*, pp. 137-8.

the sharing of hardware except in emergency planning. The fact that, as Walter McDougall puts it, American space diplomacy in this sphere in particular 'amounted at best to a benign hypocrisy' only served to illustrate how the elevation of an elaborate Happyville on the foundations of a basic Pain City remained the Rocket State's definitive structural feature.[26]

Moreover, if the space program's formal mechanisms of global integration failed to deliver James Webb's universal frontiersmen to the promised land, neither did its administrative paradigm prove the custodian of Thomas Paine's cooperative commonwealth. Webb's 'prototypes for tomorrow' were vast organizations like Exxon or General Motors or the wartime Manhattan District atomic bomb project, all powerful bodies in the tradition of Rathenau's 'cartelized state' but none of them models of popular sovereignty. Likewise, his chosen future governors were men like James Webb himself, the interchangeable elite representatives of Chicherin's administrative state but hardly an articulate *vox populi*. A career that took him from a congressional administrative assistantship during the New Deal via numerous powerful corporate and governmental posts to a variety of consultancies and trusteeships before his confirmation as NASA Chief Administrator gave Webb vast experience of large-scale organization. At no point did this require any direct electoral mandate. Equally, NASA may have been the first bureaucracy to embark on a surrealistic adventure, as Norman Mailer suggested, but it remained at base a conventional bureaucracy, defending its own patch of much sought-after turf, extending a frontier that was primarily managerial, and disseminating orders, not revolution, from above. The unity such a guidance system provided owed more to its organizational hegemony than to its Mindy Bloth-like benevolent style. The space age dominion its Chief Administrator celebrated demanded less international cooperation than global obedience. As Webb recorded privately in early 1965, the 'real question' facing Americans was not whether the United States and the Soviet Union could advance from a partnership in space towards some more general earthly *détente* but '[w]hether we can force them to cooperate by developing so much power that there

26. Theodore Sorensen, *Kennedy* (London: Hodder & Stoughton, 1965), p. 247; H.L. Nieburg, *In the Name of Science* (Chicago: Quadrangle, 1966), pp. 33-9; McDougall, p. 360.

simply isn't any alternative' (*GR* 164-5, 338, 735).[27]

The contradictions such Cummings-like thoughts encapsulated were built into the production of the Rocket State's material core. Notwithstanding the Kennedy administration's claim that the US space program gave private enterprise an opportunity to serve the public interest; notwithstanding its assertion that just as the President's personal appeal had revitalized the American 'heart-land' so his public expenditure would improve its circulation and extend its domain, the $40 billion and more pumped into the American economy over the decade following the decision to go to the moon was injected at the behest, not of the public at large but of a rather narrower community of vested interests. On the one hand the dominant elements of the American aerospace industry, on the other the potential legislative and executive beneficiaries of an expanded national space effort. Whilst the representation of Project Apollo as an enterprise drawing on the skills of the nation at large and yielding benefits for mankind as a whole remained the conventional rationale of its liberal advocates, behind these elaborate screens there persisted the motivating dreams of monopoly and conquest which fuelled the Rocket State's growth as they had fostered its birth a generation before.[28]

Of course, one of the greatest civilian public works projects ever conceived promised a congressional pork-barrel of such irresistible proportions that a number of domestic privateering operations were only to be expected. In November 1961 the earlier selection by a NASA Source Evaluation Board of Martin Marietta to construct the Apollo Command and Service Modules was overruled following lobbying by Robert Kerr (D, Okla) and George Miller (D, Cal), the Senate and House space committee chairmen, on behalf of North American Aviation and other 'interested parties' in the southwestern United States. Given private assurances of work for Miller's Downey, California, constituents and of a new NAA plant for his own Oklahoma base, Kerr successfully pressurized his former employee Webb into changing the original decision and in return took pleasure not only in helping intermediary Bobby Baker towards a fortune

---

27. Webb, pp. 34-6; Jay Holmes, *America on the Moon: The Enterprise of the Sixties* (Philadelphia: J.P. Lippincott, 1962), pp. 190-92; McDougall, pp. 388, 413.
28. Murphy, p. 364.

from a vending machine franchise inside North American's plants, but also in helping himself to the proceeds from personal land sales for the contractor's new plant outside Tulsa.[29] Likewise, the September 1961 decision to establish the Manned Spacecraft Center on a green field site south of Houston reflected the influence of Vice-President Johnson, of House Speaker Rayburn (D, Tex), and of powerful House Appropriations Committee member Albert Thomas (D, Tex), at the same time allowing the Humble Oil Company to capitalize on their 'gift' of 1,020 acres of land to Rice University for the site by engaging in large-scale real estate profiteering and property development on their surrounding West Ranch estate.[30]

Such examples could be multiplied. Yet they were significant less as instances of what were hardly unprecedented backstage political and economic intrigues than as symptoms of a more general structure of lopsided but mutual dependence linking Washington and the armaments industry at the heart of the Rocket State. Between 1957 and 1960, the aerospace sector faced mounting difficulties as new orders for the Atlas, Titan, Minuteman, and Polaris missiles failed to compensate for reductions in aeroplane procurement following the expansion of high fixed cost industrial facilities called for by the Korean War and NSC–68. North American Aviation's production of the F–86 ended in 1958, of the F–100 in 1959; Lockheed's F–80 also ended its run in 1959, while the Boeing B–52 was due for termination in 1962. Confronted by a profits squeeze and rising indebtedness, and with firms like Douglas and Lockheed registering sizeable losses, employers responded by laying off 140,000 workers over a three-year period and by mounting a concerted campaign for a major increase in government defense expenditure. In turn these industrial needs, accommodated grudgingly by Eisenhower during his last year in office, in many ways matched the diverse requirements of a new administration elected on a platform of growth. Having gained the endorsement of the aerospace trade press by

---

29. Bobby Baker (with Larry King), *Wheeling and Dealing* (New York: W.W. Norton, 1978), pp. 103-4, 169-70; Hugo Young, Bryan Silcock and Peter Dunne, *Journey to Tranquillity* (London: Jonathan Cape, 1969), pp. 143-59; McDougall, p. 374. On Kerr in particular, see Herbert S. Parmet, *JFK: The Presidency of John F. Kennedy* (1983; New York: Penguin, 1984), pp. 207-9.

30. Murphy, pp. 208-10; Young, *et al.*, pp. 161-2; Holman, p. 48; Oates, pp. 355, 374-5.

campaigning for a military Keynesian acceleration of all missile programs, during his first quarter in office Kennedy promptly approved a variety of initiatives (such as a doubling of Minuteman procurement from Boeing) to inject some $2.1 billion into an industry which was not only of strategic importance but, both as a major employer in the politically vital state of California and as a sizeable foreign exchange earner, also of electoral and economic significance.[31]

There was, however, a difficult balancing act to perform in this concert of interests. For if he was willing to stimulate the aerospace industry, and through it the economy at large, Kennedy was also concerned not to stimulate the armed forces into making ever greater and often conflicting demands for ever more expensive and often unjustifiable items. In particular he, like Eisenhower, did not want to let the US Air Force into a manned space program which was, after all, designed to express the benevolent objectives of American enterprise. Moreover, unlike his predecessor, Kennedy knew that under the leadership of the new Secretary of Defense Robert McNamara, large-scale changes in both organization and procurement were in store at the Pentagon, most notably an expansion and unification of the civilian executive staff's powers and the introduction of modern cost accounting. Such changes were designed both in general to bring the military under control and specifically to further the reorientation of the Air Force's center of gravity away from its now relatively expensive and soon to be obsolete Strategic Air Command and towards its Ballistic Missile Division. And yet initiatives like the introduction of a more competitive tendering and more carefully supervised contracts primarily threatened not the powers of the Pentagon top brass but the future profits of the industry whose support Kennedy so assiduously courted.[32]

In this context, NASA's Project Apollo offered a number of potential advantages. A decision to support a multibillion dollar expansion of the civilian space program would appease the Congress (and in particular key southern legislators whose

---

31. Desmond Ball, *Politics and Force Levels* (Berkeley: University of California Press, 1980), pp. 254-61; Charles D. Bright, *The Jet Makers* (Lawrence, Kansas: Regent's Press of Kansas: 1978), pp. 127-8, 191-2; Richard A. Smith, *Corporations in Crisis* (Garden City, N.Y.: Doubleday, 1963), ch. 3; Holman, p. 247; McDougall, pp. 321, 428-9; Young, *et al.*, p. 118.

32. McDougall, pp. 200, 321; Ball, pp. 68-79.

support the President's entire agenda depended upon) by offering its members sizeable funds to distribute with the obvious political benefits such federal largesse entailed. It would also satisfy the aerospace industry's demands for additional procurement by placing sizeable and long-term new orders with most of the major contractors. Moreover, it would do both of these things without feeding the already voracious appetites of the Pentagon's more ambitious thinkers. On the contrary, by intervening in the established relationships within the military–industrial complex the Apollo program might sufficiently weaken the link between the Pentagon and the aerospace giants to allow McNamara to press on with his reform of a more isolated service establishment. It was precisely this argument which McNamara and James Webb placed at the top of their original draft report to the President advocating the manned lunar landing in May 1961. And while the final memorandum substituted for the strategic and economic rationales the more familiar litany of scientific and prestige factors, the importance of the former was evident. If, as McDougall suggests, the specter of a moon manned by Russians – one graphically prefigured by the achievements of Yuri Gagarin – tipped Kennedy towards the conclusion that the United States 'could not *not* do it,' then, as science adviser Jerome Weisner later recalled, McNamara's confirmation that without Apollo lay-offs in the aerospace industry were imminent 'took away all the arguments against the space program.'[33]

The Defense Secretary's final point found support in one study which concluded that 'the main impact of NASA employment during the first half of the 1960s was as a stabilizing influence on aerospace employment in the Los Angeles–Long Beach–Anaheim area.' Likewise, McNamara's attempts to bring the Pentagon under greater civilian control also began showing signs of success by mid-1964. However, while such outcomes might be narrowly interpreted as vindications of civilian control in the Defense Department and of the beneficial effects of a buoyant private sector, and thus as successful expressions of the Rocket State's happier dimension, a short review of relations between NASA and its contractors in general and of McNamara's failure to have comparable reforms effectively implemented in the space agency illustrates the ways in which the manned space program

---

33. McDougall, pp. 320-22.

was anything but a tribute to the American private enterprise system – or at least the system as espoused by its promoters.[34]

James Webb's oft repeated declaration that 90 per cent or more of NASA funds were passed on to the private sector was intended to counter charges that the Apollo program merely fuelled the extension of a gigantic centralized state bureaucracy whose administrative tentacles threatened to strangle the free enterprise system. But defining a federal agency as an efficient conduit rather than as an inflexible harness did not alter the fact that, however accurate the statistic, its putative implications were either meaningless or ironic. The space agency's prime contractors may have been privately owned and managed, but the extent of their dependence on the state for survival ranged from the considerable to the total. In June 1966, North American Aviation, the recently merged McDonnell Douglas, and Grumman all had between 75 and 100 per cent of their employees engaged on government contracts; even Boeing, whose commercial aircraft sales remained buoyant for much of the 1960s, relied on NASA and the Defense Department for between a quarter and a half of its work.[35]

Correspondingly, if the first footprint on the moon defined the reach of American enterprise (as North American's publicity hand-outs implied), its achievement had little to do with any recognizably classical marketplace. Rather, as the battle for the main Apollo order demonstrated, what competition there was took place behind the screens of industrial secrecy and congressional confidentiality, with NASA distributing prime contracts according to the dictates of political and industrial equity in order somehow to preserve the 'competitive' positions of those very same favored contractors whose selection only served further to diminish the prospects of future competition. Between 1959 and 1969, only 10 per cent of the total value of NASA procurement awards were fixed price, advertised, competitive contracts, while at the industrial level the argument that large-scale state financed research and development work tended to promote industrial concentration found confirmation in the mergers of McDonnell and Douglas Aircraft in 1966, and of North American and Rockwell the following year, as well as in comprehensive studies

---

34. Holman, pp. 192, 237-8, 266; Nieburg, pp. 49, 352-68; Bright, pp. 70-73.
35. Nieburg, p. 220; McDougall, pp. 381-2; Holman, p. 260; Bright, p. 142.

of NASA's contractual data. The space agency may have awarded a large *volume* of contracts to small businesses – around two-thirds of all awards between 1963 and 1966 – but in terms of *total value* its reliance on an oligopolistic corporate elite was even more marked than the Pentagon's. NASA's four largest contractors accounted for 51 per cent of total agency procurement expenditure to June 1966; its top 100 suppliers accounted for all but 7 per cent. Four main companies directly employed only about a quarter of the total Apollo labor force. This clearly did not mean, however, that only one in four employees were dependent on those four for their welfare. On the contrary, extensive subcontracting distributed not only work but also the influence of firms like North American, a process whose boundaries were further widened as space agency budgets tempted previously non-aerospace firms into the market for federal cash.[36]

Nor was efficiency much in demand from the corporate giants operating under the space agency's generous contractual terms. During Fiscal Year 1962, some 84 per cent (and as late as Fiscal Year 1965 some 71 per cent) of the total value of NASA research and development awards were of the cost plus fixed fee variety, whilst over the same period the use of incentives to improve contractor performance was extended from a mere 2 per cent to no more than 15 per cent of the same total. Rather than imposing penalties for underbidding or inadequate performance, such contracts actually rewarded cost overruns, unnecessary elaboration in production ('gold-plating'), and internal profiteering. Not surprisingly, although the moon landing itself was completed within the time frame proposed by President Kennedy, both Mercury and Gemini projects ran well behind schedule and well over budget estimates whilst still leaving prime contractor McDonnell Aircraft unpenalized for the first job and well-rewarded with bonuses for the second. Competitive tendering and incentive contracts may have been inappropriate for such tasks, as Arnold Levine points out. But such arguments do not

36. Holman, pp. 66-7, 69-70, 77, 78, 263, 269, 280-84, 300; Barber, pp. 71-90; Nieburg, pp. 61-84; Bright, pp. 141-2, 144; Ralph Lapp, '$10 Billion More for Space?' *New Republic*, 21 February 1970, pp. 16-17; Charles E. Nathanson, 'The Militarization of the American Economy,' in David Horowitz (ed.), *Corporations and the Cold War* (New York: Monthly Review Press, 1969), pp. 207, 215-21. Cf. Walter Adams and Horace Grey, *Monopoly in America: The Government as Promoter* (New York: Macmillan, 1955).

alter the fact that, for the aerospace industry at least, the terms of much NASA contracting remained only the latest turns of the 'fixed roulette wheel' previously concealed by the Operation in the 'Forbidden Wing' of Pynchon's Casino Hermann Goering: a set of controlled conditions under which, as Levine himself acknowledges, 'the contractor could not lose (*GR* 209).'[37]

As for the space program's contribution to long-term economic growth, the *sine qua non* of the Kennedy promise and the essential drive behind the Orpheus Theater's elevator network, internal NASA studies revealed no reliable evidence to link investment in research and development with greater economic growth or productivity or the inauguration of Webb's routinely improving social system. On the one hand, there remained too many variables to establish any such connection with confidence. On the other, what indications there were suggested that whilst federal spending on research and development padded gross national product figures it had little overall effect on economic growth. It was not only critics like Seymour Melman and H.L. Nieburg who criticized such expenditure for directing skilled labor and other resources away from the producer and consumer goods industries towards the military/space sector and thereby fostering long-term unemployment and inflation. In its special space issue of June 1962 *Fortune* made a number of similar points. The American mission to the moon might vindicate the 'private enterprise system' while leaving its terrestrial powerhouses starved of resources. In practice, even as the federal government invested some $146 billion in research and development of various sorts between the flights of Sputnik I and Apollo XI, the nation itself moved towards a state of undeclared civil war, in large part as a direct consequence of the applications of such investment in southeast Asia.[38]

None of which meant that what the House Committee on Science and Astronautics defined in May 1961 as 'the ingenuity

---

37. Holman, p.78; Murphy, p. 171; Levine, pp. 74-5, 78, 93-8; Nieburg, pp. 69-71, 370; McDougall, p. 438.

38. Nieburg, pp. 64-6, 71-4, 95, 226; Amitai Etzioni, *The Moon-Doggle: Domestic and International Implications of the Space Race* (Garden City, N.Y.: Doubleday, 1964), pp. 72-112; Seymour Melman, *Our Depleted Society* (New York: Holt, Rinehart, & Winston, 1965); Gilbert Burch, 'Hitching the Economy to the Infinite,' *Fortune*, June 1962, pp. 123 ff.; Barber, pp. 24-31; McDougall, pp. 384, 437.

of the American business community' was entirely absent. On the contrary, even with the assistance of its highly placed legislative and administrative friends, the ability of the aerospace industry to convert the space agency from a supposedly responsible watchdog into little more than a profitable division of private enterprise demonstrated what was clearly an 'extraordinary capacity,' albeit not so much one 'for meeting the needs of the nation' as for meeting more narrowly defined corporate targets. This was evident in a general sense at the financial level between Fiscal Years 1961 and 1964, when the expansion of the Kennedy administration's research and development budget from $9.3 billion to $14.7 billion allowed the private sector to reduce its share of what was supposedly one of its characteristic functions – risk taking – from 35 per cent to little more than 20 per cent of the total. It was still more graphically displayed at the technical level over a similar period when, following the establishment of NASA in 1958, this new centralized body became the subject of industrial colonization.[39]

From its inception the space agency's legislative and executive overseers (not to mention its own administrative establishment) emphasized the importance of contracting out as much as possible of its work to private industry. Such out-sourcing would be in keeping, they argued, with America's faith in a free enterprise system as a superior alternative to Soviet-style state control. However, it proved difficult to afford the private sector a maximum role in the space program while simultaneously expecting it to perform according to the requirements of its federal overseer. Instead, encouraged by congressional license, the aerospace industry took the opportunity to move in on that putative sponsor. No sooner had the Jet Propulsion Laboratory at the California Institute of Technology been moved from the jurisdiction of the US Army to the NASA fold in December 1958, than it began to be stripped of its contracts and workforce. The same experience awaited von Braun's US Army Redstone Arsenal after it was reorganized as the space agency's Marshall Spaceflight Center in July 1960. Attracted by private contractors' salaries which were unconstrained by civil service pay scales even as they remained funded from federal coffers, the former

---

39. Nieburg, p. 220; McDougall, pp. 228, 383-4; Gibney and Feldman, p. 120.

Redstone establishment was converted by 1964 into a thin shell of NASA administrators surrounding a solid core of private employees bearing the insignia of Boeing, General Dynamics, IBM, and others. A comparable process was in evidence at JPL. As NASA's in-house research and development expenditure fell from 25 per cent to 10 per cent of the total R & D budget between Fiscal Years 1961 and 1963, so its share of the space program's total corps of scientists and engineers followed suit, dropping from 32.4 per cent in 1960 to 15.6 per cent in 1966. Just as a generation before the nascent American aerospace industry had used the Paperclip program to net between $400 million and $2 billion worth of missile hardware, engineering reports and drawings, and German technical expertise, so in the early 1960s it took advantage of comparable license to purchase publicly financed expertise and to eliminate a privately feared competitor.[40]

Relieved of its inherited in-house facilities, NASA proved both unwilling and unable to respond. Recommendations of the May 1962 Bell Report into federal procurement highlighting the need to raise civil service salary scales and to increase the number of positions exempted from such restrictions were implemented only slowly and never in full. Notwithstanding James Webb's claims for the potential of the new Electronics Research Center announced by NASA in 1964, the Bell Report's call for a rebuilding of in-house research and development facilities was acted on only at the Department of Defense. Comprehensive reforms introduced there by Secretary McNamara were introduced but reluctantly at the space agency, where before 1965 centralized auditing and the specter of the General Accounting Office remained distant threats to the aerospace lobby. Finally, what efforts the agency did make towards recreating independent advisory powers – most notably hiring AT&T (American Telephone and Telegraph) and General Electric in 1962 to provide technical analysis and material assistance (both on non-competitive, cost-plus contracts) – simply allowed two more large corporations to profiteer on contractual loopholes, convert

40. Nieburg, pp. 219-20, 223, 230-43; Levine, pp. 65-8, 111-12, 116; Holman, p. 66; Frederick I. Ordway III and Mitchell R. Sharpe, *The Rocket Team* (London: Heinemann, 1979), pp. 314, 349-53; Clarence Lasby, *Project Paperclip* (New York: Atheneum, 1971), pp. 26, 42-5, 251-64. Cf. Michael H. Armacost, *The Politics of Weapons Innovation* (New York: Columbia University Press, 1969), p. 155.

advisory into executive functions, and invest in new plant (particularly computers) at the expense of the federal government.[41]

But if the Rocket State's engines pressed forward on federal fuel, its crewing capacity depended on tanks which were soon to run low. With most of the basic Apollo construction work completed by 1966, and with the space agency's budget under increasing pressure as the parameters of the American Century started to come apart both at home and in southeast Asia, NASA's expenditure and its labor force began to diminish. Between the summers of 1966 and 1968 a reduction in the agency's budget of $1 billion was accompanied by lay-offs at a rate of 6,000 per month. From a peak of 420,000 in June 1965 the total work force fell steadily until by June 1971 it had reached less than 149,000, a reduction of some two-thirds in the space of five years. After completion of engine testing at the Mississippi Test Facility in 1966, the payroll fell by $18 million within three years; over the same period the Marshall Space Flight Center shed almost half of its employees, cutting its wage bill by $40 million during 1968 alone. The newly created 'humming industrial areas' of the south by no means fell silent, but it was clear that Gerhardt von Göll's dream of a terminal homeostatic control system remained predicated on parameters which were anything but homeostatic. For in reducing the structural significance of the old agrarian order this influx of federal finance simply made the south ever more dependent on aerospace spending. With reductions in the space agency's budget during the late 1960s the consequences were only too clear. Whereas in 1965 the Mississippi Test Facility had provided nearly four-fifths of Hancock County's entire non-farm employment, by 1972, with employment at the facility reduced by over 76 per cent, not only Hancock County but also Pearl River County, Missouri, Brevard County, Florida, and New Orleans, Louisiana, all featured on the US Department of Labor's list of areas with substantial unemployment.[42]

This slowing of the Orpheus Theater's elevator system in turn undermined the dream of a unified society even as the President gave voice to its realization. In 1964 Lloyd Berkner of the

---

41. Nieburg, pp. 256-66, 334-49; Levine, pp. 75-84, 112-13.
42. Levine, pp. 107, 134-8; Holmes, pp. 194, 203-4, 213, 238-41, 354-5; Murphy, pp. 392-4.

National Academy of Sciences could still advocate an expansion of scientific research and development since 'for each new Ph.D we could employ 5 to 10 engineers and for each new engineer 10 to 15 skilled workers.' Such visions of a functionally molded society bereft of class or other conflict were, by the late 1960s, exposed as fictions of the most fragile kind. If highly qualified technical staff, their numbers swelled partly by NASA's own training schemes, found themselves laid off in their thousands, unable to secure comparable work locally, and entitled to only minimum state welfare benefits as the Apollo launch program came to an end, then their prospects remained brighter than those of their putative functionaries – from construction workers to administrative support staff – whose horizons were narrower. As for the 'other Americans' whose interests the Apollo program was imagined to serve, their direct benefits proved to be even more limited. Partly because of the high proportion of well-qualified personnel employed by the space agency – 25 per cent – its labor force remained predominantly white, particularly at the highest levels. As late as 1973, little more than 3 per cent of NASA's scientists and engineers were Black. Moreover, if the coming technocracy was supposed to undermine the significance of race, it did not prevent one Black family being refused service in the dining room of the Cape Kennedy motel owned by the astronauts, nor did it stop interested parties from keeping the astronaut corps itself totally white.[43]

None of which defined some inexplicable failure. President Kennedy's expansion of the Rocket State laid the groundbase for Berkner's intimations of pan-classical collaboration, but it did so only as a result of a military Keynesianism which eased the pressure on corporate profits engendered by the deflationary strategy pursued by Eisenhower's conservative Treasury Secretary Arthur Burns during the late 1950s, and thereby contained the attendant flare-up of industrial conflict and incipient union busting. For all of the space agency's reproduction of middle-class ease, at base the labor force remained means to the Operation's ends. This was particularly evident at the recently

---

43. Lloyd Berkner, *The Scientific Age* (New Haven: Yale University Press, 1964), pp. 29-30; McDougall, p. 440; 'Growing Up With Rockets,' *Equinox*, prod. and dir. Nancy Yasecko, Channel Four Television, 9 October 1986; Holman, pp. 204-5, 238; Levine, pp. 115-21; *New York Times*, 3 June 1965, p. 21; Nieburg, p. 11.

discredited General Electric, where the securing of non-competitive, cost-plus contracts from NASA did nothing to eliminate the incentive payments scheme or to revive the cost of living adjustments respectively introduced and abolished following the unions' defeat in the 1960 General Electric strike – and it remained true across the corporate landscape. Like the astronauts, employees were defined as no more than components which added 'considerably to systems effectiveness' when given adequate instruments, controls, and training – such as the industrial relations techniques exemplified by the astronauts' plant visits.[44] While President Nixon may have been moved to declare 20 July 1969 a 'National Day of Participation,' ordering all public offices to display the flag and remain closed 'so that as many of our citizens as possible will be able to share in the significant events of the day,' his magnanimity was not universally shared. In St. Louis hundreds of AT&T employees went on strike when management declined to follow the President's lead; in California most aerospace companies likewise refused to pay their workers to sit at home watching the fruits of their labor, an attitude endorsed by both Mayor Alioto in San Francisco and the Los Angeles City Council. The National Day of Participation remained at odds with much everyday life. Even as Apollo XI made its way to the moon, Wall Street analysts were forecasting an unpromising future for aerospace stocks in the context of declining NASA expenditure, conditions which only underlined the need for the workers to stick to their tasks. And even as 8,000 Western Electric employees voted with their seats and stayed home anyway, the Houston garbage workers John Glenn so envied elected to work through the day: quite simply, they needed the cash.[45]

According to Norman Mailer, what they had missed they would not miss for long, at least not if they shared his feelings of

44. Mike Davis, *Prisoners of the American Dream* (London: Verso, 1986), pp. 121-5; Swenson, *et al.*, pp. 195, 241, 255; Mary Kaldor, *The Baroque Arsenal* (1982; London: Sphere, 1983), pp. 160-64. In 1960 General Electric was found guilty of violating federal anti-trust laws and overcharging customers following a Department of Justice investigation of price fixing in the electrical machinery industry. See Charles C. Alexander, *Holding the Line: The Eisenhower Era, 1952-1961* (Bloomington: Indiana University Press, 1975), p. 109.

45. *New York Times*, 17 July 1969, pp. 1, 22, 63, 65, 20 July 1969, sec. 3, pp. 1, 5, 22 July 1969, p. 1.

narcosis at the pre-flight news conference, of ebbing excitement within minutes of President Nixon's 'most historic telephone call,' and of mounting boredom after ninety minutes of footprints. Such feelings proved accurate. For although the flights of Apollo XII in November 1969 and of Apollo XIII the following April provided a second installment for those who had previously missed out, followed by the even more familiarly compulsive thrills of an on-board explosion, public interest dwindled thereafter. What had begun as a whirlwind romance drifted on as a marriage of dulled inconvenience; no longer 'too thrilling for words,' as one of Nixon's cabinet members described the launch of Apollo XI, but simply a drain on the nation's reserves. However inadvertently, the remarks of long-rehabilitated moon walker Pete Conrad whilst waiting to begin the laborious process of storing samples and equipment on board the lunar module of Apollo XII – 'I feel just like a guy at the shopping center with the groceries waiting for his wife' – anticipated precisely that sense of routinization which critics of manned space flight had been predicting almost thirty years before. As Mailer expected, within two years of the first lunar landing 'the murderous laws of fashion' were taking over and the viewers were looking for change.[46]

If technological novelties like thermal underwear or cordless razors fulfilled their needs, then, notwithstanding the *a priori* hollowness of such justifications, the manned space program might yet claim some credit. However, beyond such celebrated instances the promise of space travel proved less than wholesome. Studies carried out by the Denver Research Institute in 1963 and 1974 concluded that, while defining and measuring technology transfer involved a number of arbitrary judgements, innovations resulting from the manned space program were in general too sophisticated, specialized, and expensive to have extensive commercial applications. Even with space agency promotion there was only a limited industrial demand for NASA patents, and what spin-offs there were hardly possessed major significance. The most important non-NASA customer for the products of manned space flight remained the Defense Department.

---

46. Mailer, *Fire on the Moon*, pp. 37, 118-19, 288; *New York Times*, 17 July 1969, p. 21; Peter Ryan, *The Invasion of the Moon, 1957-70*, rev. and enlarged edn (Harmondsworth: Penguin, 1971), p. 163; McDougall, p. 396.

Otherwise, the greatest contribution to the American technological armory provided by the lunar landing program was technology capable of being used in a lunar landing program.[47]

Not surprisingly, therefore, Lyndon Johnson's pre-launch endorsement of James Webb's oft repeated assertion – if Americans could land on the moon then, given the willpower, they could eradicate poverty, feed the hungry, and 'do anything that needs to be done' – remained unproven. Americans were not short of determination, but they were driven in given directions. Even as a million converged on Cape Kennedy to witness the start of Apollo XI's epic voyage, thousands of Brevard County inhabitants who qualified were being denied federal food aid because the local authorities had not set aside matching funds. The $2.5 million spent by Brevard County annually on road maintenance amounted to over twice its annual welfare expenditure. As the only Black doctor in the locality remarked, 'I guess they're more concerned about promoting tourism in Brevard County than caring about hungry people.' Those figures who, like Tyrone Slothrop before them, ended up framed and unidentified – this time amidst the rumble of cars – were predictably absent from the Cape Kennedy Hilton when one airline's offer of $25,000 moon flight reservations promised to bring Barron Hilton's vision of a heavenly hotel chain one step closer to realization. If there were significant communally oriented facilities under construction, like the Bay Area Rapid Transit system in San Francisco, their existence owed more to smog and congestion than to the potentials of space age technology. As the *New York Times* lamented when difficulties continued to plague the new Penn Central Metroliner during the summer of 1969, the United States could send men to the moon but it seemed unable to make the trains run on time (*GR* 742).[48]

As for 'all those good folk' awaiting access to the Orpheus Theater, President Kennedy's promises of 'knowledge and understanding' were eventually realized, although hardly in the interests of all people. The Communications Satellite Act of August 1962 provided for the establishment of a state licensed

---

47. Holman, pp. 317-24; Nieburg, p. 78; McDougall, pp. 438, 531; Kaldor, p. 146.

48. *New York Times*, 14 July 1969, pp. 1, 23, 17 July 1969, p. 26, 21 July 1969, p. 15; *Time*, 16 July 1979, p. 17.

private monopoly – the Comsat Corporation – whose prime aims were to undercut rival Anglo-French cable systems, to extend American control of strategic world communications, and to return to the new organization's governing powers (notably AT&T, ITT, RCA, and General Telephone) sizeable profits and an expanding market, not to ennoble Fanon's wretched of the earth. Likewise, the International Telecommunications Satellite Consortium, or INTELSAT, established under the majority ownership and management of Comsat two years later, remained less the mechanism of 'world peace and closer brotherhood' which Kennedy had forecast than a profit distribution complex for the world's top industrialized countries. Since, in spite of the President's recommendations, both organizations concentrated their attentions on the more profitable parts of the globe, the first involvement most third world nations had with advanced satellite technology came courtesy of a rather longer-established institution which began operating the first truly global system of space communication in June 1966. This was the US Department of Defense, whose network relied upon two ground stations in South Vietnam and a number of others across the southern half of the globe to facilitate the mobile communications necessary for the implementation of its counterinsurgency programs. Such a system certainly brought some knowledge of the American way of life to southeast Asia, but hardly the peace and brotherhood its employers professed to defend.[49]

With the manned lunar landing program over and the Rocket State's happier backdrops beginning to unravel as the American Century's offensive petered out, the fiction of beneficial technological integration apotheosized from Tranquillity Base lost its grip. At home, as Arthur Miller had already remarked, the US Congress having paid for Apollo showed little interest in sending out multibillion dollar expeditions to investigate the more proximate orbits of Harlem or Watts. Further afield, third world leaders threw the lofty pronouncements of Soviet and American

49. Herbert Schiller, *Mass Communications and American Empire* (New York: Augustus M. Kelley, 1969), pp. 63-78, 127-46; Joseph D. Phillips, 'Economic Effects of the Cold War,' in David Horowitz (ed.), *Corporations and the Cold War* (New York: Monthly Review Press, 1969), pp. 193-7; McDougall, pp. 352-60; Levine, p. 216; Frantz Fanon, *The Wretched of the Earth*, trans. Constance Farrington (1965; Harmondsworth: Penguin, 1967).

representatives alike back in their faces, condemning the prospects of an arms race in space and questioning the motives of satellite technology users as a stagnating world economy, deteriorating trading conditions, and rising international indebtedness intensified competition for commodity markets. Robert Frost's whimsical response to a 1962 congressional enquiry into the necessity of space exploration – 'I don't know why, but it's glorious ... perhaps it will improve communications between stock exchanges' – was the uninformed remark of a man who would not live to witness the consequences of such 'improvements' in the 1980s. By contrast, Susan George's later investigations of how satellite technology was being used by the US Department of Agriculture on behalf of American grain exporters to assess competitors' potential output confirmed what many had long suspected. From its earliest days that technology was part of a larger Operation, one which in May 1964 could make the first issue of Comsat stock the most oversubscribed in Wall Street history and which a generation later could leave Guatemala, for all of its oil, the very incarnation of 'Pain City.'[50]

If this larger complex was, like the moon landing itself, an expression of nature's law, it celebrated not the ideals inscribed on the side of Apollo XI's lunar module but the liberties of a less accommodating age. The hypothesis pursued by Norman Mailer in *Of A Fire on the Moon* – 'Nazism had been an assault upon the cosmos ... was space its amputated limb, its philosophy in orbit?' – may have been impossible to confirm. But his suspicion that the astronauts themselves now contained 'some of the profound and accelerating opposites of the century' previously joined in the simultaneously primitive and technologically advanced motives of the Third Reich implied a more material plot whose divulgence would spell out at least part of such an operational link. As Tom Bower has revealed, that plot led not only from the slave labor camp at the *Mittelwerk* on which Wernher von Braun and his colleagues depended to the rockets that went to the moon, but also from the 'terminal experiments' at Dachau of which Hubertus Strughold and his fellow *Luftwaffe* aviation doctors tacitly

---

50. *New York Times*, 21 July 1969, p. 7; McDougall, pp. 356, 433-4; Nieburg, p. 14; Susan George, *Feeding the Few: Corporate Control of Food* (Washington: Institute for Policy Studies, 1979), pp. 56-9; Ince, pp. 130-33; Peter Marsh, *The Space Business* (Harmondsworth: Penguin, 1985), pp. 83-7.

approved to the simulators and spacesuits that helped train and maintain the moon walkers. Contained by its growth there survived inside the Rocket State an earlier formation which also saw history as a 'large-scale biological process' and which also sanctioned barbarism as an instinctive necessity 'deeply inscribed in man's psyche.' Between the uniform smiles elicited by a wide-spread uncritical preoccupation with the Rocket State's leading men, and the universal Oven State facilitated by the most ominously classified technologies of Brzezinsky's terminal society, there remains no more than a delta–t.[51]

51. Mailer, *Fire on the Moon*, pp. 47-8, 76; Tom Bower, *The Paperclip Conspiracy* (London: Michael Joseph, 1987), pp. 1-2, 105-24, 232-48, 254-64, 276-83; Swenson, *et al.*, pp. 34-48.

# Conclusion: The Fiery Wake

## Shutting the Water Off

Living inside the System is like riding across the country in a bus driven by a maniac bent on suicide ... you catch a glimpse of his face, his insane, committed eyes, and you remember then, for a terrible few heartbeats, that of course it will end for you all in blood, in shock, without dignity.

> — Thomas Pynchon, *Gravity's Rainbow* (1973)

[He] grew tenser as the last seconds ticked off. He scarcely breathed. He held on to a post to steady himself. For the last few seconds, he stared directly ahead and then when the announcer shouted 'Now!' and there came this tremendous burst of light followed shortly thereafter by the deep growling roar of the explosion, his face relaxed into an expression of tremendous relief. Several of the observers standing back of the shelter to watch the lighting effects were knocked flat by the blast.

> — Brigadier General Thomas F. Farrell, Alamogordo, New Mexico 05:30hrs (16 July 1945)

**conclusion.** 1. The end, close, finish, terminating, wind up ... 2. An issue, final result, outcome, upshot ... 12. The action of shutting up, enclosing, or confining. 13. A binding act ...

> — *Oxford English Dictionary*

Shortly after 4:18 p.m. Eastern Standard Time on 20 July 1969 Capsule communicator Charlie Duke in the Houston Mission Control Center reported 'lots of smiling faces in this room and all over the world' to the equally elated figures of Armstrong and Aldrin on the lunar surface and Collins in the orbiting Command Module. With the Eagle safely down at Tranquillity Base following two hours of delicate powered descent, the relief was evident. As Duke told the astronauts: 'You've got a bunch of guys about to turn blue. We're breathing again.' For some, however, relief came too late. Less than a month before this, on 27 June, *Life* magazine had recorded as its cover story the desolate underside of the Rocket State's highest stage: the anonymous instamatic faces, 'half scared and half full of bravado,' of the hundreds of young Americans killed during a recent, average, and anonymous week in the Vietnam war. The story itself 'was considered a high-water mark of *Life*'s journalism,' David Halberstam later wrote (within three years the American Century's leading standard-bearer would itself be killed off), and 'probably had more impact on antiwar feeling than any other piece of print journalism.' But its conjunction with the achievements of the Apollo XI crew was illustrative of a larger transformation in the making. For whilst a succession of moon landings between 1969 and 1972 would continue to affirm its achievements, the Rocket State's structural limitations were already being realized – only most evidently in southeast Asia – and its future replacement called up.[1]

---

1. Peter Ryan, *The Invasion of the Moon, 1957-70*, rev. and enlarged edn (Harmondsworth: Penguin, 1971), pp. 112, 114; David Halberstam, *The Powers That Be* (London: Chatto & Windus, 1979), pp. 484-5. The connections between the manned space program and the Vietnam war were multifaceted. During the 1960s, NASA was fond of extolling the virtues of its enterprises by emphasizing their gigantic scale. The Apollo program, agency officials pointed out, employed almost half a million people from every state of the union who together helped design and manufacture a spacecraft consisting of millions of state-of-the-art components to be carried into space on board a 363-foot-high rocket carrying over six million pounds of liquid oxygen, liquid hydrogen, kerosene, and hypergolic fuels for an eight-day voyage covering a total of some 500,000 miles. Similar statistic-laden indices of American power could have been – and in the case of some Pentagon boosters were – constructed for the air war in Vietnam. Between 1965 and 1969, US Air Force B-52 bombers increased the amount of high explosive and incendiary bombs dropped on all parts of North Vietnam from 315,000 tons to 1,388,000 tons annually. By this time, over 100 million pounds of defoliants and crop-destroying chemicals had destroyed millions of acres of forests and productive food-growing areas at a cost of over $100 million. Between the first and last moon landings in July 1969 and December 1972, the US Air Force dropped perhaps another 3 million tons of bombs over the whole of Indo-China, 50 per cent more than the total tonnage exploded by all branches of the US Armed Forces during the entire Second World War.

The relationship between space and Vietnam had more than comparative

The established American space program was in one sense a clear and immediate victim of the Rocket State's incipient disarticulation. President Nixon's early 1970 decision to follow the least ambitious of the three alternative options mapped out by Vice-President Agnew's recent Space Task Group report on *The Post-Apollo Space Program* – to abandon the manned mission to Mars, to cancel the last three Apollo launches, and to suspend Saturn V production and testing while reducing the Apollo Applications Program to a single Skylab – constituted part of the short-lived anti-inflation policy pursued by the new administration in the contexts of continuing expenditure in Vietnam and rising budget deficits. However, since both the limits and the legacy of the established post-war orders were inscribed in the terms of its reproduction, NASA's crowning achievement can also be viewed in other, less immediate ways: first, as a marginal cause of that deepening fragmentation, and secondly, as a notable mechanism of its attendant reconstitution.[2]

To the extent that oil price rises, inflation, and undeclared domestic and international resource wars combined with an already relatively saturated consumer durables market in the

---

dimensions, however. NASA employment began to fall off just as the commitment of US technological and manpower resources to southeast Asia started rising rapidly. As Mary Holman concludes: 'There can be little doubt that the existence of the space program increased the ease with which the United States mobilized for the Vietnam conflict. The program had brought together and maintained a large reservoir of skilled, technical and scientific manpower in the aerospace industry during the early nineteen-sixties. With the phase down in the program in early 1966, those employees separating from NASA projects increased the supply of technical personnel available to satisfy defense [sic] requirements.' See Frances Fitzgerald, *Fire in the Lake* (New York; Vintage, 1973), p. 626; George Herring, *America's Longest War: The United States and Vietnam, 1950-1975* (New York: John Wiley, 1979), p. 152; Hugh Higgins, *Vietnam*, 2nd edn (London: Heinemann, 1982), pp. 83-4; Mary A. Holman, *The Political Economy of the Space Program* (Palo Alto, Cal.: Pacific Books, 1974), p. 249.

2. Arnold S. Levine, *Managing NASA in the Apollo Era*, NASA SP-4102 (Washington, D.C.: National Aeronautics and Space Administration, 1982), pp. 260-61; Thomas P. Murphy, *Science, Geopolitics, and Federal Spending* (Lexington, Mass.: D.C. Heath, 1971), pp. 378-90; Walter A. McDougall, *The Heavens and the Earth* (New York: Basic Books, 1985), pp. 421-2; Rowland Evans and Robert D. Novak, *Nixon in the White House* (London: Davis-Poynter, 1972), pp. 177-94; Jim Campen, 'Economic Crisis and Conservative Economic Policies: US Capitalism in the 1980s,' *Radical America*, 15, nos. 1-2 (1981), pp. 37-8.

early 1970s to undermine the 'intensive regime' of wage-led mass consumption that had for thirty-five years mustered and molded the Rocket State's millions, then vast Keynesian public works projects like the space program played their part in facilitating the very extension that led to its breakdown. If only by being converted into ever rising energy consumption and ever growing demands for state financed infrastructures like roads, schools, and other services, NASA's salaries helped develop this provisionally beneficial mechanism of the Rocket State's growth to the point where it could no longer furnish the houses, cars, televisions, and suburbs which together made up the characteristic features of the post-war 'no man's land.'[3]

Simultaneously, by transferring over $17.5 billion to the southern and western United States between Fiscal Years 1962 and 1969, the space agency's budget made a proportionately greater contribution to what was the ultimately more painful mechanism, not only of the Rocket State's growth but also of its eventual succession. The contribution itself had two distinct yet related dimensions. On the one hand, the more than $13 billion worth of prime contracts and subcontracts which initiated the rapid development of the Cape Canaveral region and secured the prosperity of Houston and Greater Los Angeles industry and commerce during the 1960s indirectly helped elevate new generations of Florida real estate speculators, California construction firms, and other Sunbelt entrepreneurs and financiers: a new community of interest born of the Vietnam war boom whose relative independence from the power elite's east coast operational core would allow them to ride out the recessions of the early 1970s and underwrite the rise of the New Right during the rest of the decade.[4] On the other hand, vast prime contracts

---

3. Mike Davis, *Prisoners of the American Dream* (London: Verso, 1986), pp. 181-2, 190-92, 195-9; Murphy, pp. 183-5; Campen, pp. 38-43. The full context of this disarticulation is given in Samuel Bowles and Herbert Gitnis, 'The Crisis of Liberal Democratic Capitalism: The Case of the United States,' *Politics and Society* 11, no. 1 (1982), pp. 51-93; Michel Aglietta, 'World Capitalism in the Eighties,' *New Left Review* 136 (1982), pp. 5-35.

4. Murphy, pp. 47, 183-5; Davis, pp. 168, 171-4, 193-4; Alan Wolfe, 'Sociology, Liberalism, and the Radical Right,' *New Left Review* 128 (1981), pp. 18-19. These interests included men like Barron Hilton whose tourism and leisure fortune, considerably boosted by his successful Cape Kennedy hotel, helped finance the election campaign of Ronald Reagan. They also included firms like Houston building contractors Brown and Root, long-standing financiers and

awarded to companies like Boeing and North American Rockwell by NASA during the 1960s directly fuelled corporate giants which, while more closely aligned with the Republican establishment or pro-military Democrats than with the New Right's highest circles, were nevertheless integral to Ronald Reagan's election in 1980.[5]

Of course neither the marginal disarticulation of the intensive regime of accumulation nor the furthering of an (often overstated) shift in the geographical balance of power in the United States were part of some preordained design on the part of the aerospace industry. On the one hand, its leading constituents have always promoted the manned space program as of long-term benefit to the nation as a whole (witness the advertising campaign launched by McDonnell Douglas in support of NASA's space station during 1987). On the other, they have done so primarily to validate relatively short-term corporate accumulation strategies rather than to bolster specific political coalitions. Any analysis which seeks to portray the aerospace industry as prime mover in some putative 'Sunbelt strategy' has to take into account the fact that, after California, New York State and Boston are at the heart of the industry's growth. Nor should the importance of the aerospace/NASA nexus to the American political economy be overestimated. While the writings of Melman, Kaldor, and others illuminate the ways in which military (and space agency) Keynesianism undermines the very grounds on which it stands, whether by wasting human and material resources, distorting investment patterns, fuelling inflation, undermining economic growth, or bolstering unemployment, Project Apollo accounted for only a small percentage of federal revenues when compared, for example, with the costs of the

---

beneficiaries of Lyndon Johnson, whose securing of the multi-million dollar contract for the Manned Spacecraft Center's main complex in 1961 helped it to become one of the nation's leading open-shop constructors during the 1970s and more recently a beneficiary of Reaganite labor policy. See Stephen B. Oates, 'NASA's Manned Spacecraft Center at Houston, Texas,' *Southwestern Historical Quarterly* 67, no. 3 (1963/1964), p. 356; Davis, pp. 132-3, 138-53, 172.

5. These interests included in General Electric a firm which not only pioneered the strategy of Sunbelt industrialization and open shop employment in the late 1950s, but which also rescued Ronald Reagan from a waning Hollywood career before putting him on the road to the Sacramento state house as Governor of California in the 1960s. See Davis, pp. 121, 171-2.

Vietnam War. (The United States poured more money into the war in 1967 alone than it spent on the entire lunar landing program.) Companies like Boeing and North American Rockwell did not, in any event, compare in terms of size with the giants of the oil and automobile industries. Nevertheless, to the extent that NASA's budget swelled the largely Vietnam-related federal deficit and underwrote a standard of living dependent on continued low oil prices, it played a significant role – one, moreover, that was highly visible – in driving the Rocket State on to the point of collapse.

As for the mechanism of succession, what connected the Rocket State to its budding replacement was what united the Sunbelt's *nouveau riche* and the corporate establishment into a single community of interest: the armaments and aerospace budgets of the Department of Defense and the space agency. For whilst grassroots cadres of single issue lobbyists were mobilizing during the 1970s around anti-busing, anti-gun control, anti-abortion, and other banners to penetrate and undermine the Rocket State's liberal upper reaches, the Operation's surviving elite and its more recent recruits were coming together to press for the reseeding of that state's abiding groundbase: the largely military-related expenditure that would secure the 'hothouse economic conditions and high growth rates' in the Sunbelt upon which they both depended. In the long term that mechanism would help draw together a larger social formation of 'haves' – comprising middle-class strata of salaried managers, professionals, and qualified technicians, and those skilled white workers retained within the shrunken perimeters of the organized labor force – whose gilded lifestyles in turn depended, even in the relatively well-organized aerospace sector, on a 'split-level economy' underpinned by low wage, non-unionized employment. As the Rocket State's reproduction began to falter in the late 1960s, however, the immediate conjunction behind this larger reconstitution was the state of the aerospace industry.[6]

By the time President Nixon announced his spending cuts in early 1970, the industry had already entered a renewed depression. With military, commercial, and space agency sales all falling off from 1968 onwards, aerospace stocks reached the

---

6. Davis, pp. 169-72, 177-8, 206-30, 304-5; Allen Hunter, 'In the Wings: New Right Ideology and Organization,' *Radical America* 15, nos. 1-2 (1981), pp. 115-17, 121-38.

predicted record lows and aerospace employment continued its downward trend. A reduction in federal research and development expenditure of over $1 billion between 1967 and 1971, with cancellations not only of NASA projects but also of the US Air Force's Manned Orbiting Laboratory, led to lay-offs at McDonnell Douglas and the newly merged North American Rockwell and to a reduction in the industry's total labor force of over one-third. By June 1971 the number of workers engaged on NASA contracts in California had fallen to 35 per cent of its already shrunken June 1968 figure, leaving Los Angeles alongside the likes of Brevard County, Florida, on the US Department of Labor's list of regions with substantial unemployment. However, with the presidential election of November 1972 on the horizon, the sheer extent of the Operation's long-standing investment dictated a change. While NASA Administrator James Fletcher's November 1971 warning that the contractors could not survive another year of falling space agency orders may have been the overpessimistic assessment of an interested participant, even a depressed aerospace industry in crucial states like California, Texas, and Florida was as unwelcome to Nixon as it had been to Kennedy a decade earlier. Deliverance was correspondingly timed. Having already rescued Lockheed from the self-made scandal of the C–5A aircraft by placing the prime contract for the Trident C–4 system with its Sunnyvale Missile Systems Division in September 1971, the administration relieved the now renamed Rockwell International in July 1972 by awarding its Palmdale facility (also in California) the first $2.6 billion installment of the prime contract to build NASA's new Space Transportation System, or shuttle.[7]

As the country went to the polls that year unemployment in parts of the Los Angeles metropolitan area was falling to two-thirds of its mid-1971 level. Yet the decisions which led to this reduction – and which helped Nixon to a landslide victory – were symptomatic of a larger reconfiguration taking place behind the Rocket State's crumbling facade that would within a decade

7. Charles D. Bright, *The Jet Makers* (Lawrence, Kansas: Regent's Press of Kansas, 1978), pp. 73-4, 141, 144; Holman, pp. 236, 238, 352-4, 363; Ralph Lapp, '$10 Billion More for Space?' *New Republic*, 21 February 1970, pp. 17-18; Murphy, pp. 391-3, 482-3; James G. Phillips, 'The Lockheed Scandal,' *New Republic*, 1 August 1970, p. 23; McDougall, pp. 422-3; Mary Kaldor, *The Baroque Arsenal* (1982; London: Sphere, 1983), pp. 57-61; Malcolm McConnell, *Challenger: A Major Malfunction* (London: Simon & Schuster, 1987), pp. 44-5.

curtail the prospects of comparable new employment, not only in the aerospace sector but across the nation as a whole. For whereas the Apollo program had been nodded through Congress as a prestigious and accommodating civilian extravaganza, the space shuttle only received executive and legislative endorsement as a practical and economic instrument of military strategy. Forced to reduce its initial cost estimates by 50 per cent in order to secure political approval, NASA reduced the shuttle's payload capacity, restricted its 'fully reusable' features, and redesigned its facilities according to the Pentagon's specifications: lowering its operational envelope, for example, to render it primarily service-able for military satellite deployment and advanced weapons research. These alterations may have succeeded in getting the STS off NASA's drawing boards and into the air, but they also let open a screen that had been closed off since the agency's birth.[8]

In terms of hardware, logistics, and personnel, of course, the American space program had been intimately connected with and thoroughly dependent upon both the Defense Department and its constituent armed forces from the outset. Thus whilst the Apollo boosters were themselves custom-built, 'for everything between sounding rockets and the Saturn I NASA relied on vehicles successfully developed by the Air Force between 1954 and 1959 – notably the Atlas, Thor, and Titan ballistic missiles.' The space agency also made use of the Army Corps of Engineers for its large construction projects, of Pentagon contract adminis-tration services for its procurement supervision, of USAF facilities at Cape Kennedy and Vandenberg Air Force Base (between 1958 and 1966) for its launches, of the Defense Department's manage-ment services for its manned spaceflight tracking network, and of US Navy vessels for its capsule recovery operations. In addition, the seedbeds of NASA's manned and unmanned technological research and development, the Redstone Arsenal team and the Jet Propulsion Laboratory, were delivered from the Army's safekeeping. But these were functional, not executive ties, and whilst the Air Force had endeavored to penetrate and colonize the space agency's hierarchy from its earliest days, such attempts had been zealously challenged – particularly when James Webb was in charge. However, having proffered the Air Force a poten-

---

8. Holman, p. 354; Davis, pp. 136-53; McDougall, pp. 422-3; McConnell, pp. 32-7.

tial replacement for their previously cancelled Dyna-Soar and Manned Orbiting Laboratory systems in order to secure its own manned future in space, NASA found itself accommodating not a temporary passenger but a permanent co-pilot as the Air Force pressed its claims. Moreover, these claims, which eventually resulted in the construction of a military shuttle with its own launching facilities at Vandenberg Air Force Base, were in turn only part of a much larger operation challenging the very utility of a civilian space agency whose complex origins reached back to the start of the space age.[9]

Eisenhower's decision to freeze the US Air Force in particular out of the manned space program resulted from his immediate desire to retain control over military expenditure, but it also derived from what would become a more long-term strategy towards the uses of outer space demanding constraints on the Air Force's manned and unmanned activities. This strategy, formally adopted in August 1958 following the approval of NSC 5814/1 and subsequently affirmed by Eisenhower's successors, involved the adoption of a dual track approach consisting of two interlocking stages. The first stage was designed to boost the civilian and military objectives of the US space program into two distinct trajectories. On the one hand, the peaceful, cooperative, and scientific nature of American efforts would be promoted by the well-publicized voyages of the space agency's prestigious Mercury, Gemini, and Apollo spacecraft. On the other, its less appealing features would be concealed in the military budget. While agencies like the US Information Service, Radio Free Europe, and the Voice of America joined the commercial media in celebrating the achievements of American astronauts during the 1960s, from 1959 onwards publicity surrounding Department of Defense satellite launches was gradually curtailed until in May 1962 the Department imposed censorship on all such activities. The second stage was then intended to separate these unpublicized elements into defensive and offensive modes, inserting what were considered to be strategically stabilizing military satellites into permanent orbits but returning potentially disruptive

9. Levine, pp. 77, 79-80, 121-2, 211-13, 228-32; William H. Schauer, *The Politics of Space* (New York: Holmes & Meier, 1976), pp. 51-3; H.L. Nieburg, *In the Name of Science* (Chicago: Quadrangle, 1966), pp. 47-51; David Baker, *The Shape of Wars to Come* (1981; Feltham: Hamlyn, 1982), pp. 132-7.

hardware to the confines of Pentagon planning.[10]

The first of these stages had its greatest impact on the American public and Washington's other target audiences around the world; the second was directed towards Soviet political and military elites with the objective of securing the routinization of what was becoming a perilously impulsive military order. Potentially offensive space systems like anti-satellite weapons or orbital warheads simply threatened to extend an already explosive structure into new dimensions; passive systems – and reconnaissance satellites above all – actually promised to help stabilize the arms race on earth. By providing reliable information on Soviet military capabilities such surveillance mechanisms would give American civilian planners the ability to contain the demands of the Pentagon top brass for ever larger budgets and arsenals.[11] More important still, they held out the prospect of much faster progress on arms control by obviating the need for unpopular and unreliable on-site inspections or for provocative and illegal activities like U–2 overflights. Particularly after the Soviet Union began sending up its own photographic reconnaissance satellites in 1962, American policy sanctioned the deployment of only those instruments – for ballistic early warning, electronic or photographic intelligence, communications, and the like – which enhanced the characteristically precarious but equally imperative stability of the Rocket State's ultimate holding pattern (Mutually Assured Destruction) arrived at in early 1964.[12]

10. McDougall, pp. 178-85, 191, 272, 346-8; Erik Bernouw, *Tube of Plenty: The Evolution of American Television*, rev. edn (New York: Oxford University Press, 1982), p. 309; Thomas C. Sorensen, *The Word War: The Story of American Propaganda* (New York: Harper & Row, 1968), p. 159; Schauer, p. 54. Paul B. Stares, *Space Weapons and US Strategy: Origins and Development* (London: Croom Helm, 1985) is a comprehensive study of the development of the US military space program.

11. E.P. Thompson, 'Why is Star Wars?' in E.P. Thompson (ed.), *Star Wars* (Harmondsworth: Penguin, 1985), p. 14; Daniel Deudney, 'Unlocking Space,' *Foreign Policy* 53 (Winter 1983/84), pp. 94-5; Stares, pp. 54-8, 66-71. In June 1961 Defense Secretary McNamara used data gathered – ironically – by the joint USAF/CIA Discoverer satellite series to confirm American strategic missile superiority and thus to secure his staff from the toxins of the Strategic Air Command. See Desmond Ball, *Politics and Force Levels* (Berkeley: University of California Press, 1980), pp. 100-102; Philip J. Klass, *Secret Sentries in Space* (New York: Random House, 1971), pp. 71, 98-106.

12. McDougall, pp. 115-17, 180-81, 331-2, 335, 337; Paul Stares, 'U.S. and Soviet Military Space Programs: A Comparative Assessment,' *Daedalus* 114, no. 2 (1985), p. 133.

By contrast, any system which threatened to undermine what became an unwritten agreement between the superpowers not to proceed with the deployment of offensive military hardware in space was either cancelled or restricted to the laboratory. Thus early Pentagon initiatives in anti-satellite (ASAT) technology, fractional orbital bombardment systems, and space-based ballistic missile defenses under the umbrella of its BAMBI project in the late 1950s were ruled out during the Kennedy administration because, by sanctioning comparable Soviet developments, they indirectly threatened the security of American reconnaissance satellites. Likewise, the US Air Force satellite interceptor research project, SAINT, was downgraded and finally dropped when the Soviet Union abandoned its own short-lived campaign against spy satellites in favor of deploying its own. Following the August 1963 signature of the Partial Test Ban Treaty, which outlawed atomic testing in outer space, the mutual understanding which dictated these restraints was further codified through a series of diplomatic initiatives, most notably the Outer Space Treaty of January 1967 (which banned weapons of mass destruction from earth orbit and all weapons from the planets) and the SALT–1 and Anti-Ballistic Missile (ABM) Treaties of May 1972 (which guaranteed the security of the surveillance satellites on which their verification depended and restricted both the extension of ABM systems and the use of ASAT weapons).[13]

The overall strategy may have appeared to sanctify the American space program and to restrict at least the scope of the arms race. However, in practice it merely contained behind bipartisan screens the seeds of its own dissolution by simultaneously assuming and undermining the permanence of what proved to be no more than a temporary technological milieu. As early as August 1958, the National Security Council had recorded that the satellites being developed to monitor Soviet armaments levels might also be capable of targeting the American strategic deterrent; by the 1980s over twenty years of innovation pioneered by Heinlein's 'backbone of American industry' had led to the original division between offensive and defensive systems being all but obliterated. Not only were reconnaissance satellites functioning as integral

---

13. Alexander Flax, 'Ballistic Missile Defense: Concepts and History,' *Daedalus* 114, no. 2 (1985), p. 49; McDougall, pp. 191-2, 339, 349, 415-19, 430-31; Stares, *Space Weapons*, pp. 53, 71, 88-90, 101-5, 106-29, 158, 237-8.

parts of a nuclear arsenal whose governing strategic doctrine was itself being updated towards a first strike counterforce posture, so too were those instruments formally employed for test ban monitoring and early warning, for geodesy and navigation, and for meteorology and communications whose sheer quantity, growing strategic importance, and continuing exposure were in turn making them ever more obvious targets. Correspondingly, in the early 1960s the Russian military space program remained in its infancy; by the mid-1970s the resumption of previously intermittent anti-satellite weapons tests and the burgeoning use of space on the part of the Soviet armed forces had prompted Washington to consider abandoning what was an ostensibly rescinded agreement. Since the Russians were apparently engaged in renewed research into offensive technologies while both American military and civilian space budgets continued to fall, by 1976 there remained only political barriers to the birth of an arms race in space.[14]

The resumption of Soviet satellite interceptor tests in 1976 was only the immediate catalyst for the decision of the outgoing Ford administration to authorize the development of a new American anti-satellite system in January 1977. It was in itself not a sufficient cause. As Paul Stares has pointed out, these tests were if anything less successful than those which had brought minimal American concern when carried out by the Russians between 1968 and 1971. Ford's decision was also motivated by changes in what was perceived to be the overall balance of power between east and west and by related changes in the domestic American polity: on the one hand, the concerted effort on the part of the Soviet Union to close the military distance between itself and the United States (particularly in the field of strategic weapons) and its alleged adventurism in the third world; on the other, the mounting pressures exerted following the resignation of President Nixon in 1974 and the final defeat in Vietnam in 1975 for a comprehensive process of national 'rearmament' – to be expressed most clearly through the expansion and modernization

---

14. Robert Aldridge, *The Counterforce Syndrome* (Washington: Institute for Policy Studies, 1978), pp. 14-20; Herbert F. York, 'Nuclear Deterrence and the Military Uses of Space,' *Daedalus* 114, no. 2 (1985), pp. 20-22; Stares, *Space Weapons*, pp. 14-17, 140-46, 239-43, 252-3; Stares, 'U.S. and Soviet Military Space Programs,' pp. 128-9; Deudney, pp. 95-6; McDougall, p. 433. On Heinlein, see above p. 113.

of American military resources – following years of 'one-sided *détente*.' And as events in the Horn of Africa, Iran, and elsewhere were drawn on during the Presidency of Jimmy Carter to fuel the resulting drive towards what Fred Halliday describes as 'the second Cold War,' so the remaining constraints on the extension of the arms race into space were eroded. For while Defense Secretary Brown's December 1978 implementation of Ford's original decision was justified as a means of securing bargaining counters for simultaneous anti-satellite 'weapons control talks,' the Soviet invasion of Afghanistan one year later resulted in not only the suspension of those talks and the abandonment of *détente* in general but also the legitimization of the unrestrained military option advocated by the backers of candidate Reagan.[15]

## A Prophecy of Escape

> Our cause is based on the eternal principle of righteousness; and even though we who now lead may for the time fail the cause itself shall triumph ... We stand at Armageddon, and we battle for the Lord.
>
> — Theodore Roosevelt (August 1912)

> The Strategic Defense Initiative ... isn't about war. It's about peace. It isn't about retaliation, it's about prevention. It isn't about fear, it's about hope, and in that struggle, if you will pardon my stealing a film line, the Force is with us.
>
> — Ronald Reagan (March 1985)

Elected on a platform which berated the timidity of American foreign policy since the Vietnam war, denounced the SALT–2 agreement initialled by Jimmy Carter and Leonid Brezhnev in June 1979, and proclaimed the existence of a strategic 'window of vulnerability' that proved to be as fictional as the 'missile gap' of

---

15. Stares, *Space Weapons*, pp. 19, 135-40, 158, 168-72, 177-99, 237-43; Stares, 'U.S. and Soviet Military Space Program,' pp. 134-5; Fred Halliday, *The Making of the Second Cold War* (London: Verso, 1983), pp. 76-7, 97-104, 122-5, 149-58, 214-33; Michael T. Klare, *Beyond the 'Vietnam Syndrome'* (Washington: Institute for Policy Studies, 1981), pp. 1-14; Noam Chomsky, *Towards A New Cold War* (London: Sinclair Browne, 1982), pp. 1-57, 188-229; Jerry Sanders, *Peddlars of Crisis: The Committee on the Present Danger and the Politics of Containment* (London: Pluto, 1983), pp. 149-315.

an earlier generation, Ronald Reagan furnished the executive joint around which the Operation's established and insurgent investors could negotiate their mutual advance. Within little more than two years of the new President's inauguration the central terms of that advance had been provisionally inked in: at the practical level, a $1,600-billion five-year military budget that would by 1983 increase the projected expenditure of the Carter administration by almost 20 per cent and thereby finance an acceleration of the comprehensive modernization program that had first been conceived as the Rocket State started to falter; at the theoretical level, a virtually formal commitment to a first strike counterforce war-winning capability that would by the same year be encoded not only in Reagan's description of the doctrine of Mutually Assured Destruction as 'a ridiculous plan' but also in his related enthusiasm for the latest lines in survivalist technology: the hardware of 'strategic defense.'[16]

These aggregate terms were promptly translated into material and tactical form. On the one hand, an expanded budget allowed the new administration to push forward with the foundations of its latest theater of war. Within a week of his inauguration President Reagan approved contracts worth over $400 million for ASAT research and development. Between August and October 1981 the White House announced its decision to order the development of hardware for a space-based ABM system and rejected Soviet draft proposals for a space weapons treaty whilst showing little or no interest in presenting counterproposals for future discussion. During 1982 the Pentagon claimed that the Soviet Union could be in a position to deploy a functioning orbital laser weapon within a year, the General Accounting Office released an independent study calling for expenditure sufficient to develop an American equivalent for testing by 1990, and the Defense Department's space budget surpassed NASA's for the first time since 1960. On the other hand, these budgets facilitated the long-awaited take-off of the theater's main load-bearing structure, the space shuttle, whose successful introduc-

---

16. Halliday, pp. 52-5, 70-75, 123, 234-8; Robert Scheer, *With Enough Shovels: Reagan, Bush and Nuclear War* (New York: Random House, 1982), pp. 8-13, 66-82, 233-4, 250-51; Klare, pp. 10-12; Thompson, 'Why is Star Wars?' pp. 16-19; Peter Pringle and William Arkin, *SIOP* (London: Sphere, 1983), pp. 190-96; Fred Kaplan, *Dubious Specter* (Washington: Institute for Policy Studies, 1980), pp. 40-54.

tion enabled its promoters to start distributing their latest productions. On the completion of the shuttle's maiden voyage in April 1981, the US Air Force opened a Space Laser Office; within two months of the orbiter's first all-military mission in July 1982, that office was absorbed by the new Air Force Space Command established at Colorado Springs to coordinate the now much widened range of high altitude military activities. Subsequently, by deploying a variety of Defense Department satellites essential to the execution of the Reagan administration's updated nuclear war plan, SIOP–6, and by providing an orbital base for testing the ASAT lasers, ABM particle beams, and space-to-ground weapons proposed by the various service branches, the shuttles performed a number of backstage operations before the forthcoming sell-out's main feature.[17]

Rehearsed and rigged up behind the Rocket State's longstanding screens, these fortified leading lights took the stage as that once padded shell came apart. Where the 'inverse mapping' witnessed by Tyrone Slothrop amidst the rubble of Berlin in 1945 had relocated overt militarism inside the folds of civilian adjustment, now that armored inheritance reasserted itself in the hearts of resurgent Americans, the homelands of war-ravaged neighbors, and – most ominously – the previously inviolable haven of space. Rather than resurfacing in the Oven State's guise, however, it kept on its frayed but still comforting cover. When, on 23 March 1983, President Reagan unveiled to the world the ultimate act of a long-running epic – the dream of strategic defense – he promised not Hitlerite blood and struggle but 'new hope for our children,' not vengeful slaughter but salvation for all mankind. During the following weeks and months, as his proposal entered its post-launch boost phase, it was presented not as a skylight to war but as both an ethical alternative to the 'unthinkable' morality of frozen deterrence and a reliable assurance of

---

17. Stares, *Space Weapons*, pp. 217-30 passim, 258-9; Walter A. McDougall, 'How *Not* to Think about Space Lasers,' *National Review*, 13 May 1983, p. 552; Baker, pp. 137-43, 154, 164-5; E.P. Thompson, 'Folly's Comet,' in E.P. Thompson (ed.), *Star Wars* (Harmondsworth: Penguin, 1985), p. 161, n. 95; William Arkin, 'Why SIOP–6?' *Bulletin of the Atomic Scientists*, April 1983, p. 10; Pringle and Arkin, pp. 175-6, 185-8; Peter Marsh, *The Space Business* (Harmondsworth: Penguin, 1985), p. 114; Halliday, p. 239; *New York Times*, 22 June 1982, p. A19, 29 June 1982, pp. A1, C3.

survival beneath a caring and calming 'peace shield' (*GR* 372-3).[18]

This reassuring tableau presented the lineaments of the Rocket State's replacement. On the one hand, by anticipating the construction of a shelter whose very passivity and invulnerability sanctioned its employment come the shadow of disaster, the Strategic Defense Initiative projected a fortified joint command structure at the intersections of security and liberty, of innocence and power, and thereby previewed the eventual reintegration of the post-war order's now overextended and overtaxed network of load-bearing catches. It offered a fiction of ultimate personal salvation – of what one critic described as being 'snatched away before the bomb' – for an audience in flight from the diminishing holds of a now profaned former life; a dream of evacuation beneath that latest 'cycle of infection and death' – the arc of the falling rocket – which Blicero had inaugurated in the ruins of the Oven State's former realm. On the other hand, by drawing on ever greater quantities of finance, labor, and equipment in the course of that shelter's construction, the Strategic Defense Initiative simultaneously installed a more centralized resource distribution system and thereby began to haul back into line the Rocket State's now labyrinthine elevator network. It engineered a way out from the terminal pointlessness of an overcrᵒ vded Happyville by directing its passengers towards the more disciplined attractions of a now refurbished and bloodless Pain City.[19]

These evacuations hardly constituted new departures, however, since both the reopened route and its recently ordained destination were drawn up within the long-term schedules of the Operation and its augmented elite. Just as the absorption of militarism in the ruins of occupied Berlin was only ever a reassignment of roles designed to carry the Operation across and beyond 'the awful interface of V–E Day,' so its resurrection before the flagging shrouds of an abandoned liberalism marked no more than the latest recasting of Pynchon's 'nonstop revue' in the interests of its established directors. Boeing had produced

---

18. Thompson, 'Why is Star Wars?' pp. 23-6; Thompson, 'Folly's Comet,' pp. 93-7; Stares, *Space Weapons*, p. 225; Alan Bullock, *Hitler: A Study in Tyranny*, rev. edn (Harmondsworth: Penguin, 1962), p. 398; McDougall, 'Lasers,' p. 550; York, p. 27.

19. Marlene Tufts, 'Snatched Away Before the Bomb: Rapture Believers in the 1980s,' (unpublished Ph.D dissertation, University of Hawaii, 1986); Ronnie Dugger, 'Reagan's Apocalypse Now,' *Manchester Guardian*, 21 April 1984, p. 19.

both the first stage of the Saturn V rocket that propelled Apollo XI to the lunar surface and the B–52 bombers that pounded much of a distant southeast Asia to dust; now, under SDI, they carried out research on infrared sensors and kinetic energy ballistic missile defenses. TRW Corporation had formerly manufactured both the descent engine for the Apollo lunar modules and the American soldier's M–14 rifle; now they studied hydrogen fluoride lasers and ABM tracking systems. In the long term, having already persuaded an amenable President to earmark some $90 billion for systems research and development over the decade following his 1984 re-election, the SDI lobby foresaw a final deployment budget of between $400 billion and $1,000 billion or more not only dwarfing and replacing established medium term contracts for the MX, Midgetman, and Trident missile complexes but also securing the health of the aerospace and electronics industries well into the twenty-first century.[20]

Yet while such well-ordered junctions of security and escape were established along the Operation's projected main lines, they promised rather less than a standardized service. Whereas the manned space program had been an integral part of a politico-economic strategy designed to accommodate and transform previously unreconstructed domains like the south, the Strategic Defense Initiative stood poised to become the central element in a rather more direct campaign to preserve a coalition of mostly Sunbelt-centered 'haves' against the claims of intentionally stranded 'have-nots.' Indeed, what Matthew Rothschild and Keenen Peck foresaw as a burgeoning 'Star Wars society' – characterized by the comprehensive militarization of scientific research, industrial production, and academic life – prefigured the apotheosis of Mike Davis' ascendant split-level America: on the one hand, a gilded and sensitive safehaven of 'sumptuary

---

20. John A. Adam and Mark A. Fischetti, 'SDI: The Grand Experiment,' *IEEE Spectrum*, September 1985, p. 57; Stares, *Space Weapons*, p. 215; Nieburg, p. 212; Thompson, 'Folly's Comet,' pp. 131-4; Ralph Lapp, *The Weapons Culture* (New York: W.W. Norton, 1968), pp. 186-7, 192-3. Of the total contracts for research into SDI and ASAT technology awarded in Fiscal Years 1983 and 1984, some 87 per cent went to 10 contractors, 8 of which were included in the Defense Department's 20 largest arms suppliers. These 10 included most of NASA's manned space program prime contractors, from Rockwell (formerly North American Rockwell) and Boeing to McDonnell Douglas and Lockheed. For the consequences for aerospace stocks see *New Republic*, 12 August 1985, p. 33.

suburbs and gentrified neighborhoods' held together within foundations of secure, often arms-related employment and 'superstructures of social liberalism' and 'occupied by the middle classes, the rich, and elements of the skilled white working class'; on the other, the correspondingly demonized outlands of deindustrialized ghettoes reserved for the multiplying contingents of those low-waged and non-unionized, unwaged and insecure under-classes on which the Sunbelt's selected reposed.[21]

This rising order was, in fact, less than entirely novel. Even as the expansion of the aerospace industry into the south under the aegis of NASA drove the Rocket State towards its climax during the 1960s its successors' lineaments were already being sketched out in the offices of vast west coast aerospace giants like Boeing and North American. For while the US space program's growth fostered the creation of uncounted thousands of jobs for all types of workers, from painters to personnel managers, from tour guides to trainee astronauts, it also entailed an additional emphasis on high technology production which in turn implied a restructuration of the industry's core productive workforce. 'The triumph of the squares,' as one space agency official described the Apollo VIII mission of 1968, therefore had implications extending far beyond the walls of NASA's Mission Control Center. Whether involved in the development of specialized alloys or of state of the art computer programs, the growing number of well-qualified, technically sophisticated, middle-class workers such production involved – 'the guys with crew-cuts and slide rules who read the Bible and get things done' – adumbrated the formation of a new coalition that would eventually gain leverage over the nation's entire political economy.[22]

In one, albeit simplified, sense, this transformation can be associated with the displacement of combat aircraft as the leading lights of US military strategy by succeeding generations of ballistic missiles. As early as 1954, the year of the von Neumann report on ICBM feasibility, the total number of aircraft produced in the United States under the joint stimulus of NSC–68 and the Korean War had begun a sustained, decade-long fall, only new

21. Matthew Rothschild and Keenen Peck, 'Star Wars: The Final Solution,' *The Progressive*, July 1985, pp. 20-26; Thompson, 'Folly's Comet,' p. 129; Davis, pp. 206-21, 226-30, 242-3, 304-5.

22. *The Times* (London), 10 January 1969, p. 8. I am indebted to Mike Sprinker and Mike Davis for their discussions of this point.

government orders for Atlas, Titan, Polaris, and Minuteman missile systems in the late 1950s holding off a deeper aerospace depression. In 1960, when a Cornell graduate named Thomas Pynchon arrived at Boeing's Seattle plant to study handling techniques for the Bomarc guided missile and to publish a short paper summarizing his conclusions, John F. Kennedy was already campaigning for high office on a platform that emphasized the need for an acceleration of US missile production. The following year, the new President doubled the number of Minutemen ordered from Boeing under the previous administration's procurement program and further committed the United States to a military strategy centered on the US Air Force's Ballistic Missile Division and the Navy's Polaris submarine fleet at the expense of the Strategic Air Command. In 1962, when Pynchon left Boeing, production of the B–52 came to an end and his former employers proceeded to concentrate on Minuteman, a short range attack missile (SRAM), and the first stage of the Saturn V launcher that would later carry the Apollo spacecraft to the moon. The overall movement was obvious. At the end of World War II, 'Rocketman' Tyrone Slothrop had been 'the wave of the future.' Fifteen years later it was the turn of the rockets themselves.[23]

Both the industry's conversion from aircraft manufacture to aerospace research and development and the consequences of this shift for the labor force were recorded by Pynchon in his first two novels. In *V.*, published the year after Pynchon left Boeing, Clayton Chiclitz' Yoyodyne Inc. grows into a post-war aerospace giant manufacturing gyroscopes for aircraft and missiles and supplying telemetry equipment, electronics components, and communications gear, eventually developing by the end of the 1950s into 'an interlocking kingdom' producing a wide variety of high technology systems inside the burgeoning military–industrial complex.[24] Three years later, in *The Crying of Lot 49*, the firm's Galactronics Division extends its operations directly into the missile and spacecraft business. In the process it not only becomes the main source of employment in the surrounding community of San Narciso, Southern California, but also takes

---

23. Bright, pp. 127-9, 140-41; Thomas Pynchon, 'Togetherness,' *Aerospace Safety*, December 1960, pp. 6-8. On the aerospace industry in the late 1950s, see above, pp. 215-17.

24. Thomas Pynchon, *V.* (London: Jonathan Cape, 1963), pp. 226-7.

on the characteristic features of an elite white collar domain far removed from the wartime world of Rosie the Riveter and her male successors in airframe assembly: air-conditioned offices full of IBM typewriters, fluorescent bulbs, and swivel chairs on which well-trained electronics engineers gyrate to consult 'fat reference manuals' or to refold 'rattling blueprints.' That its annual stockholders' meetings are attended by an 'invasion of Yoyodyne workers' underlines the politically orthodox nature of these favored employees like the engineers whose working patterns allow them 'to avoid responsibility' and whose income patterns dispose them, according to the company song at least, to 'swear undying loyalty' to their employers. Again, we are a long way from the conditions of the great 1941 North American Aviation strike. And if Pynchon provides little more than a glimpse of the nascent split-level society's lower echelons, where electronics assembly workers get drunk in a local bar rather than attend the stockholders' meeting, he does emphasize that between the corporate headquarters, shopping malls, and housing tracts of San Narciso lies 'the plinth course of capital on which everything afterward had been built, however rickety or grotesque, towards the sky.' San Narciso, as its name implies, is a breeding ground for a culture in love with a dream image of itself. It, too, is a wave of the future. Twenty years later, it would be part of a much larger world of hypnotic and material elevation, lavishly endowed by the nation's most famous Californian, whose prime beneficiaries would seek almost literally to pave over the burgeoning poverty traps of East Los Angeles in order to provide themselves with more long-term ideal homes.[25]

To many Americans – whose numbers by 1980 included some forty million born-again Christians – the very exclusivity of these divisions actually had its attractions. The barely concealed fantasies of imminent deliverance encoded in such diverse recent Hollywood productions as Steven Spielberg's *Close Encounters of the Third Kind* (1977) and John Milius's *Red Dawn* (1984) rehearsed dreams of personal elevation for a public subjected to rapid technological change and incessant military alerts without reference to social context beyond moments of utopian harmony

25. Thomas Pynchon, *The Crying of Lot 49* (1966; New York: Bantam, 1967), pp. 12-14, 30-31, 59-62, 84; Mike Davis, '*Chinatown*, Part Two? The "Internationalization" of Downtown Los Angeles,' *New Left Review* 164 (1987), pp. 79-81.

or anarchic vengeance, and proved to be as profitable as the apocalyptic and equally groundless dramas worked out by Rudolph Maté in *When Worlds Collide* (1951) and Alfred Green in *Invasion USA* (1953). Of course, that these films' immediate material contexts actually required such dehistoricization (for all of the consciously labored allusiveness of a Milius) was itself an expression of the new order's precarious state. Twenty years earlier, at the height of the Rocket State's parabola, more politically realistic fictions of Armageddon could still be blithely accommodated. In response to the computerized detonation of the Doomsday Machine in Stanley Kubrick's *Dr. Strangelove* (1963), the protagonist refuses to 'rule out the chances of preserving a nucleus of human specimens' – naturally including 'our top government and military men' – at 'the bottom of some of our deeper mine shafts.' Since the survival of an elite prepared 'to foster and impart the required principles of leadership and tradition' would be 'absolutely vital,' Strangelove continues, any popular opposition to such a move could legitimately and successfully be dealt with by armed force. 'Men cannot fight against tanks and machine guns, Mister President,' he concludes, 'this we have proved.' In the 1980s, such assertions would define the politics of vengeance and its contingency planning rather too plainly.[26]

But if at the peak of the Rocket State's rise in July 1969 NASA theoretician John Houbolt could imagine that 'the world ought to stop right at this moment,' as that state's prospects turn ever downwards during the 1980s his project could yet be pressed home. For retreat to the shelter of an armored, resolute, and self-possessed Star Wars society is a fiction of security that, once realized, would make disaster the ultimate fact. As a comprehensive body of political, strategic, and economic studies have shown, SDI's official claims cannot be met. Even as the entire system appears technologically impractical it remains strategi-

---

26. Halliday, p. 114; Scott Forsyth, 'Evil Empire: Spectacle and Imperialism in Hollywood,' in Ralph Miliband, Leo Panitch, and John Saville, (eds) *Socialist Register 1987* (London: Merlin Press, 1987), pp. 105-6; Bill Warren, *Keep Watching the Skies! American Science Fiction Movies of the Fifties. Volume 1, 1950-1957* (Jefferson, N. Cal. and London: McFarland, 1982), pp. 58-65, 77-9; Peter George, *Dr. Strangelove* (London: Corgi, 1963), pp. 139-43. On British emergency regulations, see Duncan Campbell, *War Plan UK*, rev. edn (London: Paladin, 1983). I know of no comparable study of American planning.

cally destabilizing, financially insatiable, and, in any event, practically untestable. Rather than fostering disarmament it promises to undermine remaining prospects for overall weapons control and to fuel the armaments race at the strategic level. Instead of promoting *détente*, its very development seems predicated on a drive for pre-emptive attack. As Nobel laureate Hans Bethe wrote in late 1984, 'it is difficult to imagine a system more likely to induce catastrophe.'[27] In his futuristic manifesto on the potentials of space exploration issued on the eve of the first manned lunar landing, NASA Administrator Thomas O. Paine predicted that 'American vision and determination' would by 1984 be helping 'to sweep aside old world dogmas, prejudices, outworn traditions, and offensive ideologies.' But that summer, as the United States successfully tested a kinetic energy ABM over the Pacific and a laser reflector installed on the shuttle, a rather different sort of 'fiery wake' capable of sweeping aside all this and more was being pressed forward by a less celebrated 'frontier brotherhood' encamped in the laboratories of Los Alamos, Lawrence Livermore, and elsewhere. Whether this brotherhood's revolutionary efforts 'to tame its underdeveloped planet for posterity' would prove sufficiently successful to allow the offspring of Dr. Strangelove's subterranean elite a second coming in 'a spirit of bold curiosity for the adventure ahead' remained a moot point. If NORAD, the North American

---

27. Gene Farmer and Dora Jane Hamblin, *First on the Moon* (London: Michael Joseph, 1970), p. 243. The literature on SDI is already enormous: a recent bibliographical guide – Robert M. Lawrence, *The Strategic Defense Initiative Bibliography and Reference Guide* (Boulder, Colo.: Westview Press, 1986) – cites over 1,000 references. My understanding is based on the following: Hans Bethe, Richard Garwin, Kurt Gottfried and Henry W. Kendall, 'Space-Based Ballistic Missile Defenses,' *Scientific American*, October 1984, pp. 37-47; Richard L. Garwin, Kurt Gottfried and Donald L. Hafner, 'Anti-Satellite Weapons,' *Scientific American*, June 1984, pp. 27-37; Kosta Tsipis, 'Laser Weapons,' *Scientific American*, December 1981, pp. 35-41; Ashton B. Carter, *Directed Energy Missile Defense in Space* (Washington, D.C.: Congressional Office of Technology Assessment, 1984); Union of Concerned Scientists, *The Myth of Star Wars*, ed. John Tirman (New York: Vintage, 1984); David Andelman, 'Space Wars,' *Foreign Policy* 44 (1984), pp. 94-106; McGeorge Bundy, George F. Kennan, Robert McNamara and Gerard Smith, 'The President's Choice: Star Wars or Arms Control,' *Foreign Affairs* 63 (1984/85), pp. 264-78; William E. Burrows, 'Ballistic Missile Defense: The Illusion of Security,' *Foreign Affairs* 62 (1984), pp. 843-56; E.P. Thompson (ed.), *Star Wars* (Harmondsworth: Penguin, 1985). A large collection of essays are published in *Daedalus* 114, nos. 2 and 3 (1985), which are given over entirely to the debate on space weapons.

Defense Command buried deep in the granite of Cheyenne Mountain, Colorado, was no longer secure, where would future General Cummings go instead?[28]

## An Agonized Raving

Scores of students at Concord High School were gathered in the auditorium this morning, wearing party hats and cheering and blowing into noise makers as the space shuttle Challenger roared into the sky with a social studies teacher from the school aboard. They cheered more when they saw a flash . . .

'We weren't sure what was going on,' said Marsha Bailey, a sophomore, who was watching on one of the dozens of television sets installed around the school for the occasion, 'I thought it was part of the staging.'

— *New York Times* (29 January 1986)

I've gone through the simulations, but I still can't imagine it.

— Christa McAuliffe (January 1986)

One answer was placed under the spotlight that same summer when, even as backstage detachments of President Reagan's frontier brotherhood pressed on with their rehearsals for the total security state, less exhaustive auditions for a starring role in the Operation's most recent evacuation drama finally began. In 1969, and for a number of years thereafter, Norman Mailer's mill worker could only dream of becoming the ordinary American who one day might be prepared, in Mailer's words, to speak directly to God. Following the successful return of the final Apollo XVII mission a week before Christmas 1972, continued restrictions on the space agency's budget limited opportunities for manned space flight to the Skylab series, which placed a total of nine men in extended earth orbit between May 1973 and February 1974, and the Apollo–Soyuz Test Program, a politically motivated link-up which boosted three Americans and two Russians into orbit for a few days during July 1975. In August 1977, however, the first unpowered approach and landing test

28. *New York Times*, 17 July 1969, p. 47; Adam and Fischetti, pp. 34, 47; George, p. 144; William J. Broad, *Star Warriors. The Weaponry of Space: Reagan's Young Scientists* (1985; London: Faber, 1986); Pringle and Arkin, pp. 62, 167-8.

mission of the space shuttle *Enterprise* brought the prospects of access to space for the mill workers of the world visibly closer. For whereas all previous space flight had been of an exploratory nature and dominated by former military test pilots and, to a lesser degree, trained scientists, the new Space Transportation System was designed with the express purpose of advancing from exploration to exploitation of the heavens, a move which, its advocates argued, depended on NASA's making access to space a more routine and risk-free operation for the public at large.[29]

That purpose remained the formal rationale of the shuttle following the maiden voyage of the *Columbia* which lifted off from Cape Kennedy on 12 April 1981, the twentieth anniversary of Yuri Gagarin's first orbit of the earth. On 4 July 1982 President Reagan joined a half-million strong Independence Day crowd gathered at Edwards Air Force Base, California, to witness the return from its fourth and final test flight of this latest gem of the newest ocean and went on to declare that, with the completion of its proving series, the shuttle's first priority was to become 'fully operational and cost-effective in providing routine access to space.' Little more than two years later, on 27 August 1984, the President provided further convincing evidence of such an objective when he announced that he was 'directing NASA to begin a search ... to choose ... the first citizen passenger in the history of our space program.' Their search, he added, would be restricted to the national elementary and secondary school system: the first ordinary American in space was to be 'one of America's finest: a teacher.' For some two million Americans, if not for the original mill worker, the auditions which in 1969 had seemed little more than science fiction were now within reach. The following July Vice-President George Bush revealed to White House reporters that from some 11,000 applicants the space agency had selected one candidate for the flight on board the space shuttle *Challenger*: a high school teacher from Concord, New Hampshire, named Christa McAuliffe.[30]

Like an earlier generation of astronauts, Christa McAuliffe promptly found herself the object of numerous investment strate-

---

29. Norman Mailer, *Of A Fire on the Moon* (Boston: 1970; New York: New American Library, 1971), pp. 60-61; Marsh, p. 153; Baker, p. 133.

30. *New York Times*, 5 July 1982, pp. 1, 14, 28 August 1984, pp. A1, C3, 20 July 1985, pp. 1, 28; Richard S. Lewis, *The Voyages of Columbia* (New York: Columbia University Press, 1985); *Time*, 19 July 1982, p. 35.

gies mapped out by the White House, NASA, television reporters, editors, journalists, and politicians. On the day her selection was announced, space agency administrator James Beggs had emphasized that the teacher from New Hampshire was 'going to be a hot property for us,' and for the next six months, as Malcolm McConnell later confirmed, 'she was photographed, videotaped, interviewed, and carted around parade routes and conventions almost continuously.' In Concord, local dignitaries made their chosen daughter the center of attention in a celebratory parade and declared a 'Christa McAuliffe Day' in her honor. The front pages of the New Hampshire press carried 'Christa's beaming face ... countless times, floating weightless during training, dwarfed by *Challenger* before an earlier launch, grinning with her husband, Steve.' At the national level her opinions were sought on a variety of issues, particularly those relating to education policy, and her endorsement solicited – unsuccessfully – for various political causes. From the front page of the *New York Times* to the personalities page of *Time* magazine, Christa McAuliffe found herself being maneuvered across the same 'upper reaches of American protocol' that the Mercury astronauts had negotiated more than a generation before. Like those much molded stars, she was news.[31]

Not surprisingly, however, given the extent to which the Operation's priorities had changed during the intervening years, the messsage delivered by way of the first citizen astronaut was rather different than that encoded in the lives of the first men in space. Whereas astronauts like John Glenn had been explicitly celebrated as elite figures whose combat experience and aerial expertise made them self-evident heroes, Christa McAuliffe was boosted as the embodiment of ordinary America at its best: living proof, not only that access to space was becoming a reality for all, but that the qualities which had made men like Shepard and Glenn heroes could be located on any American block. If the astronauts of John Glenn's era had displayed characteristic attributes of loyalty and industry, of skill and efficiency, of piety, responsibility, and modesty in their military and space agency careers then, according to the biographical sketches which

31. McConnell, p. 102-3; *Newsweek*, 10 February 1986, p. 26; *New York Times*, 21 July 1985, sec. 1, p. 14; *Time*, 29 July 1985, p. 55; Tom Wolfe, *The Right Stuff* (London: Jonathan Cape, 1979), pp. 273-5, 358-62.

accompanied press reports of her selection, so too did Christa McAuliffe in civilian life. She was born into a typical American family and experienced an all-American childhood, taking piano and dancing lessons, joining the Girl Scouts, pitching for an all-star softball team, attending Marion High School, Framlington, Massachusetts (where, aged 15, she met her future husband), and then graduating from Framlington State College before entering the Maryland public school system as a teacher. Having helped finance her husband's studies at Georgetown University Law School, taken a Master's degree in education at Bowie State University, and moved to New Hampshire, she then turned to raising two young children while simultaneously teaching economics, law, and history at Concord High School and taking part in numerous other activities ranging from catechism class instruction to volleyball and voluntary work. No less than Alan Shepard or John Glenn, Christa McAuliffe offered 'a morally instructive parable of state.' Not only did her status as a wife, mother, and teacher provide social reassurance of America's public and private character in an abstract sense, but her self-evidently model and fulfilling life demonstrated the practical possibility of achievement for ordinary Americans of all complexions. As the editor of the *Concord Monitor* put it: 'The people in our city saw in her the best that we have to offer.' She was the worker from the education mill who had proved that it could still be done.[32]

More important still, from the point of view of her promoters, she was the worker whose achievements validated Rothschild and Peck's burgeoning 'Star Wars society': the personal embodiment and institutional advocate of a putative spiritual renaissance which had been specifically associated with that society's most celebrated material icon, the space shuttle, ever since the occasion of its maiden voyage only three months after the inauguration of President Reagan. Where the orbiter was a powerful and majestic product of daring scientific ability, free enterprise innovation, and bold political leadership which had risen like a patriotic phoenix from the ashes of a generation's military defeat, social disorder, and administrative corruption, Christa McAuliffe

---

32. *Newsweek*, 10 February 1986, pp. 24-6; *New York Times*, 20 July 1985, p. 28, 29 January 1986, p. 6; *Manchester Guardian*, 20 January 1986, p. 1; Clive Bush, *The Dream of Reason* (London: Edward Arnold, 1977), pp. 32, 41; Guy Debord, *Society of the Spectacle* (Detroit: Black & Red, 1977), para. 60.

was a figure of talent, personal initiative, and commitment whose hard work and sacrifice during that era of American stagnation had given rise to individual achievement and satisfaction. Where the orbiter was the harbinger of a high technology revolution that would bring prosperity and stability to enterprising Americans, Christa McAuliffe was an example of the well-educated middle-class professional whose role would be to introduce the citizens of the future to the vistas of space age opportunity facilitated by science, mathematics, and individual excellence. To the editor of the *Concord Monitor*, Christa McAuliffe 'stood for what was best in us at a time when we wanted to believe that the American spirit was reborn.' To the President, she and the shuttle she would ride were part of an 'investment in the future of America's greatness in space and in the young men and women of this nation who will be the leaders of tomorrow.' To a nervous middle class primed to abandon the Rocket State's faltering precepts, she rehearsed a fiction of personal deliverance within the nascent divisions of a newly disciplined, split-level land.[33]

For her part, Christa McAuliffe secured and then defined her personal elevation within the appropriate terms of inherited public mythology. 'John Kennedy inspired me with his words about placing a man on the moon,' she wrote in her application for the 'Teacher in Space' project, 'and I still remember on a cloudy, rainy night driving through Pennsylvania and hearing the news that the astronauts had landed safely.' In a longer perspective she compared herself to those pioneering women who travelled across the Plains on Conestoga wagons, keeping personal journals of their experiences and writing letters back east. She too would keep a diary in order to provide future historians with valuable documentary evidence of the experience of the final frontier for ordinary Americans in the late twentieth century. Her 'ultimate field trip' would thus serve to illuminate for the children of today the continuing desirability of the kind of personal commitments which President Kennedy had felt the lunar landing demanded in her own childhood, a message that would be beamed into schools across the United States via the Public Broadcasting System during the *Challenger*'s week in space. As she told White House reporters after Vice-President Bush had

---

33. McConnell, pp. 8, 21-3; *New York Times*, 24 August 1984, p. A1; *Newsweek*, 10 February 1986, p. 26.

announced her selection: 'I think the students will say that an ordinary person is contributing to history, and if they can make that connection, they are going to get excited about history and about the future.'[34]

This emphasis on the experience and activities of ordinary people as essential components of history, which she promoted both inside the classroom and beyond, was in theory a view diametrically opposed to that held by the President she claimed had inspired her, who had seen history as the domain of an elite and the role of the citizenry as that of an attentive but essentially passive stage crew. In practice, however, Christa McAuliffe's own contribution to history fell well within the bounds of what the late John F. Kennedy would have considered fitting since, even as it was promoted in the Operation's most recently characteristic language of part sentimental, part militant populism, it was made in accordance with terms set down from above. On the one hand, the decision to send a teacher into space came from President Reagan himself and was announced with predictable fanfare as part of an election campaign in which a number of polls had shown an otherwise popular President to be particularly vulnerable on the question of his perceived commitment to public education. On the other, the nationwide selection program thereby established gained the enthusiastic support of a space agency whose leading figures remained concerned about the declining public interest that had accompanied the shuttle's successful routinization of space travel: it was only the latest in a series of public relations exercises which had already placed two politically important Congressmen in orbit and which was scheduled to carry the first American journalist into orbit in September 1986. Before her flight Christa McAuliffe had joked to a fellow teacher that much of the training she had received 'consisted of stern admonitions to "never touch these switches."' In the final two hours before *Challenger*'s lift-off on 28 January 1986 she remained as silent as Gottfried, the sacrificial victim encased by Blicero in the 00000 rocket at the conclusion of *Gravity's Rainbow*. The astronauts may have come some distance from the 'lab rat' days at Lovelace clinic, but for an ordinary American

34. *Manchester Guardian*, 29 January 1986, p. 1; *New York Times*, 29 January 1986, p. 6; *Time*, 29 July 1985, p. 55; *Newsweek*, 10 February 1986, p. 26.

citizen wired up to the space shuttle's circuits the controls stayed a long way off (*GR* 749-51).[35]

As it happened, of course, the first ordinary American in space had no time to record her experience of space travel for posterity. Nor, for that matter, was any personal account strictly necessary. Seventy-three seconds into the flight, some four miles down range of the same Launch Pad 39 complex at Cape Kennedy from which Apollo XI had begun its voyage to the moon over fifteen years before, and nearly ten miles above the earth's surface, *Challenger* was blown apart by an explosion resulting from a catastrophic structural failure in one of the shuttle's two external solid fuel boosters.[36] The 'ultimate field trip' was over. And yet, even as the sudden destruction of one of the country's celestial flagships brought a halt to the teacher in space program, the death of the former's chosen occupants provided simultaneous impetus for the survival of the latter's interior drive. Five days before the delayed launch, the official arrival of the seven shuttle crew members at the Cape had led one press photographer to compare them with 'a war movie propaganda platoon' in their celebration of national unity. Like the service groups recorded in Frank Capra's US War Office *Why We Fight* series between 1942 and

---

35. *New York Times*, 20 July 1985, p. 28, 28 August 1984, pp. A1, C3; McConnell, pp. 7, 14, 87, 92-3, 100-103, 238; *Newsweek*, 10 February 1986, p. 21; *Manchester Guardian*, 29 January 1986, p. 8; Wolfe, pp. 79-109. The two Congressmen were Senator Jake Garn (R, Utah), Chairman of the Senate Appropriations Committee's sub-committee on space, who flew on board *Discovery* in April 1985, and Representative Bill Nelson (D, Fla), Chairman of the House Space Science and Applications Committee's sub-committee on NASA appropriations, who went into orbit early in January 1986 on board the *Columbia*. After President Reagan had announced the competition to choose a teacher, NASA administrator James Beggs added that future citizen astronauts would be selected 'from all areas of American life' and suggested specifically that 'someone from the labor movement' – perhaps 'a worker on the line' – might be chosen in order to motivate other workers to greater efforts. Given the extent of the Reagan administration's attack on the organized labor movement, this would have been a somewhat unappetizing carrot, for laid off automobile workers if not for overworked Rockwell engineers. See Davis, pp. 138-53; Kim Moody, 'Reagan, the Business Agenda, and the Collapse of Labor,' in *Socialist Register 1987*, ed. Ralph Miliband, Leo Panitch, and John Saville (London: Merlin Press, 1987), pp. 153-76.

36. For summaries and analyses of the interim conclusions of the official Rogers Commission report into the *Challenger* explosion, see *Time*, 10 March 1986, pp. 40-42; *New Scientist*, 12 June 1986, pp. 17-18, 11 September 1986, pp. 52-6. McConnell's study provides further detailed evidence.

1945, they had 'one of everything' and, as Tom Wolfe later wrote, 'seemed to have been chosen to represent Everybody's America. One was black, one was Jewish, one was Asian-American from a new state, Hawaii, and two were women.' Little more than a minute after the launch such mythical communities were inaugurated in their more familiar form as television network ratings mounted rapidly in response to the explosion. Kurt Vonnegut was right. Not for the first time in American history, personal sacrifice was the inspiration for national unity: the Kennedy promise lived on.[37]

Predictably, the official expressions of that unity, and some of the ways in which Christa McAuliffe's contribution to history actually *began* at the point of her death, were registered in full on the front page of the *New York Times* the following morning as columnists recorded how 'the nation came together yesterday in a moment of disaster and loss,' how Christa McAuliffe 'captured [the] imaginations' of Americans everywhere, and how one of millions of other average citizens had been made to feel 'like I was a part of it' by the space agency's enlistment of 'ordinary people.' No less than in the days of George Washington, death allowed the canonization process to advance unhindered, even if it meant the canonization of the routine. And yet beneath such conventional pieties, less penitent manifestations of the sense of identity promoted by this latest public sacrifice for national security came briefly to light. At Concord High School, students, staff members, and friends of the nation's first citizen astronaut who had gathered round television sets to cheer as the *Challenger* roared upwards, 'cheered more when they saw a flash' and rais[ed] their thumbs in a signal of victory' in the belief that the explosion 'was part of the staging,' a reaction also in evidence at the launch pad. In Ohio, an ex-Marine responded to questions concerning the wisdom of continuing to send civilians into space in spite of the evident risks by stating that he was 'all for civilians.' Across the country, an opinion poll carried out for *US News and World Report* revealed that some 75 per cent of Americans agreed

---

37. McConnell, pp. 93-4; *Newsweek*, 10 February 1986, p. 24; *New York Times*, 30 January 1986, p. C20. Whilst television ratings rose quickly, 'several stations logged complaints ... from viewers who wanted their favourite soap operas to be shown instead of shuttle coverage.' As one television station receptionist in Salt Lake City remarked, 'people said they were tired of watching space shuttle coverage all day – they wanted their entertainment.'

with the suggestion that the deaths of the astronauts were 'a regrettable disaster but nevertheless a price we [sic] must be willing to pay for the exploration and mastery of space.' If the benedictions heaped upon the heads of the Apollo XI crew at Cape Kennedy in July 1969 were repeated verbatim by eye witnesses to the last flight of the *Challenger*, then the woman who found it thrilling that the Saturn V rocket taking Neil Armstrong, Buzz Aldrin, and Michael Collins to the moon might explode was certainly less than unique. As an editorial in the *New York Times* underlined, 'far from turning the public against the space program, the loss of the space shuttle *Challenger* appears to have revived interest and enthusiasm.'[38]

The Presidential commission investigating the immediate causes of the *Challenger* explosion published its final report in June 1986 and concluded, in the words of Chairman William Rogers, that the decision-making process at NASA showed 'a serious deficiency' and was 'clearly flawed.' According to the assessment of one insider during the public hearings held by the Commission in early March, the accident 'was an absolutely preventable thing [which] should never have happened.' As independent studies produced both before and after publication of the official report suggested, however, the procedures which allowed the space agency to launch the *Challenger* on 28 January 1986 were but the proximate bureaucratic expressions of more far-reaching political, commercial, technical, and financial necessities whose intersection revealed that the O-ring seals ensuring the structural integrity of the space shuttle's solid fuel boosters were not the only joints to fail under pressure.[39]

The original space shuttle program had only gained political backing during the early 1970s after top NASA officials knowingly provided the Nixon administration with overoptimistic estimates for its research, development, and operating costs. A decade and more later, as the limits of the Rocket State that had necessitated such measures became increasingly evident and took on additional, novel forms, the space agency found itself commit-

38. *New York Times*, 29 January 1986, pp. 1, 3, 31 January 1986, p. A30; Bush, p. 42; *US News and World Report*, 10 February 1986, p. 21.
39. *Time*, 10 March 1986, pp. 40-42; McConnell, pp. 116, 178, 240-41. The Rogers Commission Report was published as *Report of the Presidential Commission on the Space Shuttle Challenger Accident* (Washington: US Government Printing Office, 1986), 5 vols.

ted to an ever more ambitious flight schedule in order to retain its residual support. In attempting to accommodate the diverse requirements of potential military, commercial, and scientific clients in a single, manned, reusable vehicle on a budget which, in real terms, had remained fixed from 1974 onwards at little more than one-third of its 1965 peak, the agency suffered a variety of financial and technical problems that left the shuttle program some two-and-a-half years behind schedule and 20 per cent over budget by the time of *Columbia*'s maiden voyage. While difficulties were only to be expected in such an advanced experimental endeavor, they meant that NASA's original cost estimates for shuttle launches multiplied by a factor of 10 to 15 over as many years. As costs mounted so the European Space Agency's unmanned vehicle, Ariane, began offering serious competition for the growing international commercial satellite launching business: undercutting shuttle rates with the benefit of continuing state assistance even as the US Congress sought to reduce NASA's own launch subsidies, and gaining perhaps one-third of a market worth some $500 million a year even as Washington planned to increase American prices by almost 100 per cent in 1986. Only by increasing the shuttle's flight rate could the space agency hope to cut unit costs, regain the initiative against mounting foreign competition, and retain the support of congressional budget committees already searching for ways of reducing a runaway federal deficit.[40]

However, achieving administrator Beggs' 1981 goal of twenty-four shuttle missions every year with a complement of only four orbiters rather than the five they had originally anticipated proved beyond the capacity of the space agency. Malcolm McConnell's study of the *Challenger* explosion records how from 1984 onwards workers, technicians, and managers at the Kennedy Space Center were subjected to ever longer hours and often weeks or months of unbroken working days in order to meet deadlines dictated by the increasing frequency of shuttle missions. As budgetary demands encouraged contract administrators Lockheed to improve short-term cost effectiveness by reducing the number of quality control and safety inspectors, so

40. *Science*, 13 June 1986, p. 1335; *Newsweek*, 10 February 1986, p. 20; *US News and World Report*, 10 February 1986, pp. 18-19; *New Scientist*, 26 May 1986, p. 26; McConnell, pp. 62-4; Stares, *Space Weapons*, pp. 258-9.

fewer staff carried out a rising number of post-flight inspections and refurbishment, repair, assembly, and flight certification operations on as many as three separate orbiters in a number of different buildings, activities further complicated by repeated modifications requests and a shortage of spare parts which necessitated the cannibalization of one returning shuttle in order to prepare another for flight. Consequent stress and fatigue sometimes bordering on exhaustion 'produced errors in judgements and an eventual breakdown in safety standards' which in turn led to further delays.[41]

While the *Challenger* that lifted off on 28 January 1986 was prepared under such conditions, incorporating a number of vital components removed from the returning *Columbia* orbiter only a week before its scheduled launch date, lax safety procedures and overwork amongst Kennedy Space Center staff were not the immediate causes of its destruction. In two respects, however, the pace of work demanded by a flight schedule that had converted the Shuttle Operations and Check-out Facility into what one Kennedy engineer described as 'ulcer city' did contribute to the in-flight explosion. On the one hand, it intensified rivalry between NASA's main operational centers and led senior management at the Marshall Space Flight Center in particular to impose increasingly dictatorial labor conditions, to direct staff that 'under no circumstances [would] the Marshall Center ... be the cause for delaying a launch,' and to refuse to inform either agency headquarters in Washington or other NASA centers of technical problems that might undermine their reputation for efficiency. The most persistent of these problems involved the O–ring seals securing the field joints of the shuttle's solid fuel boosters.[42] On the other hand, it made the manufacturers of those boosters, Morton Thiokol's Wasatch Aerospace Division in Utah, equally reluctant to admit that the potentially catastrophic joint rotation problem of which they had been aware since 1977 remained unresolved. Having secured a cost plus fixed fee contract for the boosters worth some $1 billion in 1973, Morton Thiokol were in 1985 negotiating the second phase of the contract with NASA to provide the additional sixty pairs of rockets called for by the accelerating shuttle program. The contract was vital to

---

41. McConnell, pp. 66-75.
42. McConnell, pp. 24-6, 72, 106-23.

the company's highly profitable Aerospace Division and prompted Thiokol management to shelve requests from some of their engineers that they inform NASA headquarters of the extent of the dangers involved in continuing to fly. Consequently, as the Rogers Commission reported in some detail, when Morton Thiokol and Marshall Center managers discussed the reservations of Thiokol engineers concerning the performance of the unreliable O–ring seals in what were record cold temperatures at Cape Kennedy the evening before *Challenger*'s flight of 28 January 1986, they ignored conventional NASA safety procedures and placed the onus of proof that the solid fuel boosters were unsafe on the engineers. Since absolute proof could only result from a booster failure in flight, the launch went ahead as planned.[43]

When Christa McAuliffe and the other six astronauts stepped on board the *Challenger* the following morning, therefore, they stepped inside a history of partisan maneuverings on the part of space agency officials, American industrialists, and the executive branch of the US government. Elevated within one of the Operation's power structures, they were destroyed within another. But the contribution of an ordinary American volunteer and her fellow astronauts could hardly be recorded as such, nor was it. Even as death removed the *Challenger* crew from that perilous history, it transferred them into the safe keeping of those fictions of elite deliverance first collected in Tom Wolfe's chronicles of *The Right Stuff*: fictions which had started to impregnate the astronauts as they edged their way onto the cover of *Life* in the late 1950s and which continued to be added to and drawn on by politicians, journalists, and other interested parties ever since.

---

43. McConnell, pp. 118, 176-201; *Time*, 10 March 1986, pp. 40-42. During 1985, as Morton Thiokol engineers concluded from studies of used shuttle boosters that catastrophic O–ring failures had already been avoided only narrowly, Thiokol contract negotiators faced an external challenge from commercial rivals who had been defeated in the original tender competition in 1973. That autumn, Aerojet General, United Technologies Corporation, and other solid fuel rocket manufacturers successfully persuaded eighty Congressmen to lobby NASA to consider tenders from other enterprises for the second stage of the booster contract which the Utah firm had until then hoped to monopolize. Faced with renewed competition, Morton Thiokol were particularly reluctant to expose persistent technical problems to congressional scrutiny. On the circumstances surrounding the original contract decision-making and Morton Thiokol's extensive Utah-based political support in both the space agency and Congress, see McConnell, pp. 51-60.

A few days before the scheduled launch of the *Challenger*, mission specialist Ronald McNair had told reporters that 'you can only become a winner if you are willing to walk over the edge.' In accordance with the long-standing maxims of such a faith, the contribution to history of McNair, McAuliffe, and the rest of the *Challenger* crew was restructured as the necessary apotheosis of a terminal society whose perfection lay just beyond reach. In his televised address to the nation on the evening of the explosion, President Reagan paraphrased the poem of a younger, earlier victim of aerial warfare in promising that 'we will never forget them, nor the last time we saw them this morning as they prepared for their journey and waved goodbye and "slipped the surly bonds of earth to touch the face of God."' In adding its own proud salute two weeks later, *Newsweek* compounded the President's words by translating the *Challenger*'s brief elevation into a quick frozen, permanent frame: having died 'in the service of their country,' all seven astronauts could 'live forever in our memories, poised in the clear blue sky almost ten miles up and climbing, in that perfect instant before holocaust.'[44]

## The Rush Hour: Here Comes Everybody

The future doesn't belong to the fainthearted. It belongs to the brave. The *Challenger* crew were pulling us into the future, and we will continue to follow them.

— Ronald Reagan (28 January 1986)

Are these the words of the all-powerful boards and syndicates of the earth? ... *Listen*: Their Garden of Delights is a terminal sewer ... All that they offer is a screen to cover retreat from the colony they have so disgracefully mismanaged. To cover travel arrangements so they will never have to pay the constituents they have betrayed and sold out. Once these arrangements are complete they will blow the place up behind them.

— William Burroughs, *Nova Express* (1966)

The morning after the *Challenger*'s destruction, Christa McAuliffe's *New York Times* obituary included a quotation in

---

44. *Newsweek*, 10 February 1986, pp. 8, 13; *New York Times*, 29 January 1986, p. 7.

which she imagined her selection for the teacher in space program as part of that larger putative dynamic of emancipation in American history perhaps most cogently recorded in the celebrated closing lines of the Emma Lazarus poem inscribed 100 years beforehand on the base of the Statue of Liberty. 'I think just opening up the door, having this ordinary person fly,' she had said in evaluating her selection, 'says a lot for the future.'[45] Not surprisingly, the prospective liberties to be bestowed by many shuttle missions on present and future generations would prove to be as hallucinatory as those anticipated during the flight of Apollo XI in 1969. The Houston contractor who in the wake of the *Challenger* explosion told reporters 'these people are heroes, dead or alive: they go out and try to make life easier for us through their efforts' may have said all the right things. The former California test pilot who had once dismissed the Mercury program as work fit only for monkeys but who now reminded the public that 'progress is marked by great smoking holes in the ground' certainly defined one familiar sanction for waste. But the Assistant Secretary of Defense who confirmed that the Pentagon would remain the biggest customer for the remaining orbiters only underlined the sort of freedom the *Challenger* and its sister ships primarily served. Having devoted its first ten pages the day after the explosion to events at the Cape, the next page of the *New York Times* reminded readers that 'MISSILE SHIELD PROGRAM GETS PENTAGON'S HIGHEST PRIORITY.' The astronauts may have gone, but the Strategic Defense Initiative in which the shuttles played an essential part lived on.[46]

The door which opened up as a result of the January explosion would not lead to greater civilian participation in or control of the exploration and exploitation of space, and had in any case been unlocked some fifteen years earlier when the space agency first took on board Defense Department specifications in order to ensure the development of the shuttle fleet. For with the suspension of all further orbiter missions pending the conclusion of internal and public enquiries into the loss of the *Challenger*, the influence over American space operations which the US Air

45. John Higham, *Strangers In The Land*, 2nd edn (New York: Atheneum, 1978), pp. 23, 63; *New York Times*, 29 January 1986, p. 5.

46. *New York Times*, 29 January 1986, pp. 1, 3, 11; 30 January 1986, p. A18; *US News and World Report*, 10 February 1986, p. 20.

Force had so doggedly pursued was noticeably strengthened. On the one hand, the halt in shuttle operations, the loss of 25 per cent of total fleet capacity, the erosion of commercial confidence, and the anticipated increase in charges for both shuttle space and satellite insurance combined to direct a greater proportion of potential commercial clients towards the alternative facilities provided not only by the European Space Agency but also by subsidized Chinese and Japanese launchers. Having written off much of the hundreds of millions of dollars it had expected to earn from commercial satellite launches between 1986 and 1990 at a time when the Grumm–Rudman–Hollings budget-balancing legislation was making a real increase in its congressional funding most unlikely, NASA found itself more thoroughly committed to the Defense Department than ever before. On the other hand, the cessation of shuttle flights and the consequences of failures sustained by American Atlas and Delta launchers during the spring of 1986 combined to ensure that the Defense Department would in turn take the initiative in plotting the future direction of its pilot civilian agency.[47]

Whereas commercial enterprises such as Western Union and Comsat could turn to European or Asian organizations for launch facilities, the Pentagon remained heavily dependent on the shuttle. As a result, even though the US Air Force had already reacted to the latter's unreliability in 1984 by gaining congressional approval for the procurement of ten new Titan launch vehicles (and would subsequently request additional conventional boosters), the grounding of all orbiters ensured that by the time flights began again a growing backlog of military payloads – over twenty by early 1988 – would be awaiting insertion into orbit. Moreover, with only three shuttles in service, such a backlog was expected to rise regardless of the frequency of their missions.[48] Under such conditions the Pentagon exercised an option granted it by President Reagan in 1982 allowing the military to displace scientific and commercial payloads from any

---

47. *Newsweek*, 10 February 1986, p. 22; *US News and World Report*, 10 February 1986, p. 16, 16 June 1986, pp. 14-15; *New Scientist*, 2 October 1986, p. 54, 22 May 1986, pp. 26-7; *New York Times*, 30 June 1986, pp. A18-19; *New Scientist*, 10 September 1987, p. 32.

48. *US News and World Report*, 10 February 1986, p. 16; *Science*, 13 June 1986, p. 1335; McConnell, p. 64.

shuttle manifest in favor of national security-related items such as intelligence satellites and SDI technology. The President's approval of this military request ensured that commercial traffic would from then on only be carried selectively and as a third priority, thereby further reducing the space agency's appeal to one of its intended prime constituencies, and that the Defense Department would consequently increase its share of the available number of flights. When NASA made public its first post-*Challenger* flight schedule in October 1986 it confirmed that military payloads would absorb four of the first five shuttle missions and that thereafter the Pentagon's overall share of launches would rise from its previous one-third to 41 per cent. Not surprisingly, the new orbiter finally approved by President Reagan in August 1986 was to be delivered to the Vandenberg Air Force Base shuttle facility in California rather than to Cape Kennedy. As the *New Scientist* concluded two months later: 'Military requirements dictated a redesign of the shuttle in the first place. Now ... it appears that military interest is the essential glue holding together what is supposed to be a civilian program.'[49]

If such decisions meant that NASA had effectively made a decision to 'paint the shuttle Air Force blue,' as one critic implied, then in the process they effectively closed off the Rocket State's final escape valve, the way out to Lyndon Johnson's ultimate vantage point beyond even the granite bunkers of Cheyenne Mountain, for the foreseeable future. In spite of the continuing (well-publicized) enthusiasm of many school children for an opportunity to fly into space in the wake of the *Challenger*'s loss, the Ilse Pöklers of the late twentieth century faced an unpromising wait.[50] In any event, as Thomas Pynchon implies through his allusions to the writings of Tsiolkovsky and Zamyatin, Franz and Ilse Pökler's dreams of reaching some state of tranquillity beyond the planet earth were never more than figments of their imagination. However distant from the cramped conditions of the original evacuation at the beginning of *Gravity's Rainbow*, and however free from the pull of gravity, such an insular high society remained chained by terrestrial bounds. Not only was the

49, *Science*, 14 February 1986, p. 666; *New Scientist*, 2 October 1986, pp. 54-5, 9 October 1986, p. 15.

50. *US News and World Report*, 10 February 1986, p. 24, 16 June 1986, p. 14; Doris Kearns, *Lyndon Johnson and the American Dream* (London: André Deutsch, 1976), p. 145.

'mathematically faultless happiness' they sought facilitated by the products of an earthly 'United State' dependent on automatism and torture for its survival: so too was their prospective Garden of Delights drawn up on untenable grounds: on the moon the oceans were barren and in the skies 'there was nothing to breathe.' Notwithstanding the claims of those futurist thinkers who saw in the promise of space travel a shining gateway that would place even Emma Lazarus' golden door in the shade; notwithstanding the imagination of the nation's first citizen astronaut who a century later took with her on board the *Challenger* a small lacquered plaque inscribed with her oft-repeated classroom motto of 'reach for the stars,' the stars lit no viable way out. On the contrary, as Pynchon makes clear when the Oven State's last clearing caves in and as Christa McAuliffe discovered to her cost, such dreams of transcendence rose on the very instruments of death they eschewed.[51]

Now in the Rocket State's last clearing, to which the most recent evacuation has brought us, those instruments are prepared to begin their final descent. As the fans reach for the latest leading lights inside the Orpheus Theater so the catches in a once binding contract come loose. With the holding pattern set out in *New Dope* breaking down completely so a 'bright angel' appears to illuminate what had long been the Operation's most carefully concealed closing frame. But in that last 'wind-beat moment' the onset of the terminal rush hour ensures that the waking has all come too late. The lonely crowd, looking back to recover their pleasure dome's long-treasured features, only find themselves turning a key to the Kingdom of Heaven. Now that the sirens have sounded there can be no more evasive action, no last minute resurrections, no special dispensations. Superman has long gone,

---

51. Konstantin Tsiolkovsky, 'Dreams of Earth and Sky,' in *The Call of the Cosmos*, ed. V. Dutt, trans. D. Myshne, *et al.* (Moscow: Foreign Languages Publishing House, 1963), pp. 81-2; James Billington, *The Icon and the Axe* (1966; New York: Alfred A. Knopf, 1970), pp. 509-11; Freeman Dyson, 'Human Consequences of the Exploration of Space,' in Eugene Rabinowitch and Richard S. Lewis (eds), *Men in Space* (1969; Aylesbury: Medical and Technical Publishing Co., 1970), pp. 21-7; Carl Sagan, *Other Worlds* (New York: Bantam, 1975); Lewis Mumford, *The Myth of the Machine. Vol. II. The Pentagon of Power* (London: Secker & Warburg, 1971), pp. 305-11; Mircia Eliade, *Myths, Dreams and Mysteries*, trans. Philip Mairet (1960; Glasgow: Collins, 1968), pp. 102-10; Ernst Nolte, *Three Faces of Fascism*, trans. Leila Vennewitz (1966; New York: New American Library, 1969), pp. 566-7.

while the Lone Ranger and his celluloid allies 'kicked upstairs' as they would be, have their own looking glass to consider. And in any event, as Pynchon puts it: 'No one was ever going to take the trouble to save *you*, old fellow.' In the last seconds, with the rocket's own passover nearing conclusion, the theater's final curtain brings the show to the end of the line. After a long and unusual marriage of convenience, the story of *High Society* falls apart as another old flame comes back for the most consuming of all one night stands. Where Blicero had once retreated to his own 'clotted islands,' this island earth now turns its face to glimpse the promised end.[52]

But the break-up is not without issue. Poised in the '[p]assage-ways of routine' between time and no time, another evacuation proceeds apace; amplified beneath the sweeping shadow of the last delta–t, the birth scream of the post-Rocket State spring gathers strength. In August 1987, no more than a decade after record numbers of Americans began queueing their way into *Star Wars*, the Congressional Record Service released a report suggesting that launch costs *alone* for a functioning Strategic Defense Initiative system could reach as much as a $1,000 billion. The birth scream of the new order heralds a Sunbelt-centered boom that would deliver a confident new society preserved for life beneath the graphic charms of Los Alamos lasers and Lockheed optics, Raytheon radar, TRW tracking systems, and other such inventions to be raised in silent and invisible session. Something similar had happened forty years before when President Truman declared the highly classified products of the Manhattan Project 'the greatest thing in history.' Such an analogy is only a beginning, however, for 'there is nothing to compare it to now.' What may follow of this latest masquerade remains a moot point yet. In July 1945 Truman received word at Potsdam of the first successful testing of an atomic device from the New Mexico desert via the following message: 'Operated on this morning. Diagnosis not yet complete but results seem satisfactory and already exceed expectations.' Now, even as the children are left to await the

---

52. On 'Looking Glass,' the Strategic Air Command's flying nuclear command post, see Pringle and Arkin, plate 7. For one larger context, see Richard Slotkin, *Regeneration through Violence. The Mythology of the American Frontier, 1600-1860* (Middleton, Conn.: Wesleyan University Press, 1973), pp.104-11.

return of God's Peace under the same basic training their forebears had received at the Rocket State's birth ('Lie and wait, lie still and be quiet'), the Operation's rising stars start to light out for territories new. As Dr. Strangelove bawls at the end of his struggle: '*Mein Führer*, I can walk!'[53]

53. George Lucas, dir., *Star Wars*, Twentieth Century Fox, 1977; *New York Times*, 2 August 1987, p. 23; Harry S. Truman, *Memoirs, Vol. 1. Year of Decisions, 1945* (London: Hodder & Stoughton, 1955), p. 352; Stanley Kubrick, dir., *Dr. Strangelove, or: How I Learned to Stop Worrying and Love the Bomb*, with Peter Sellers and George C. Scott, Columbia Pictures, 1963; Campbell, pp. 59-62, 104-11, plate 6; Peter Pringle and James Spigelman, *The Nuclear Barons* (1981; London: Sphere, 1983), p. 41. See also Norman Mailer's film treatment, 'The Last Night,' in his *Cannibals and Christians* (1966; London: Sphere, 1969), pp. 425-43, and Robert Jay Lifton, *Death in Life: The Survivors of Hiroshima* (New York: 1967; Harmondsworth: Penguin, 1971, pp. 25-6.

# Appendix 1

## Totalitarianism

Hannah Arendt's *The Origins of Totalitarianism* (1951; 2nd edn 1958; 3rd edn 1966; new edn New York: Harcourt Brace Jovanovich, 1973) was completed in 1949, four years after Hitler's death and four years before Stalin's. Its prime concern is therefore necessarily with these early totalitarian regimes rather than with the systems which developed from them. In retrospect, it is perhaps more useful to interpret Nazism and Stalinism as transitional forms of control and to view World War II – along with Pynchon – as the central phase of a transformation within the developing global system of power which now incorporates both capitalist and state capitalist societies. Arendt herself gives the end of British rule in India in 1947 as the terminus of classical European colonial imperialism, and 1930 (in the case of Stalin) or 1938 (Hitler) as the points at which totalitarian control gained full power. The intervening years may be considered as a period of reorganization in which the struggle between totalitarian and non-totalitarian powers resulted in the emergence of a hybrid. Arendt foresaw this in the late 1940s when she reasoned that 'totalitarian solutions may well survive the fall of totalitarian regimes.' In the conclusion to the second (1958) edition of *Origins* she added that:

> the crisis of our century ... is no mere threat from the outside, no mere result of some aggressive foreign policy of either Germany or Russia ... it will no more disappear with the death of Stalin than it disappeared with the fall of Nazi Germany. It may even be that the true predicaments of our time will assume their authentic form – though not necessarily the cruelest – only when totalitarianism has become a thing of the past. (p. 460)

Most recently, in her preface to the new 1966 edition, Arendt identifies anti-communism as America's own official 'counter-ideology' which mirrors many of the characteristics of those total-itarian ideologies the United States officially opposes. To this extent her analysis of totalitarianism remains applicable beyond its Nazi and Stalinist forms.

However, Arendt's analysis focuses on the radical differences between imperialism and totalitarianism and pays less attention to the fact (although she acknowledges it) that imperialism is not so much replaced by totalitarianism as absorbed into it, some-thing which is clear in the societies that emerged after World War II, for both the Soviet Union and the United States retain imperial systems and are commonly called neo- or quasi-imperialist states. Indeed, Nazi Germany itself remained an imperial power, aggressively expanding its dominion through the use of the *Wehrmacht* and reliant on an economy of looting whether capturing Rumanian oil wells, the French chemical industry, or cheap eastern European labor. In addition, the fact that the Nazi machine fell into the hands of non-overtly ideological technocrats such as Speer and Bormann towards the end of the war is the best instance of the process of succession from transitional totalitarian-ism as discussed above. For the technocratic dimension of Nazism, see the chapters on Bormann, Speer, and Heydrich in Joachim Fest, *The Faces of the Third Reich*, trans. Michael Bullock (New York: Ace Books, 1970). The persistence of imperialism as a structure of power is discussed with respect to the United States in two works by Harry Magdoff: *The Age of Imperialism* (New York: Monthly Review Press, 1969), and *Imperialism* (New York: Monthly Review Press, 1978).

As the preceding discussion suggests, there is no single satis-factory term for the international power structure which has developed since 1945. Bertram Gross, discussing characteristics of contemporary America which also apply to both western Europe and Japan, uses the phrase 'friendly fascism' in his *Friendly Fascism: The New Face of Power in America* (New York: Evans, 1980). But many of the characteristics he identifies apply equally to the state capitalist societies of the Soviet bloc. The Frankfurt School and other European scholars use the term 'late capital-ism,' as for example in the writings of Jürgen Habermas, Claus Offe, and Ernst Mandel. Other phrases, such as the corporate state or advanced industrial society, all express aspects of the system. Norman Mailer, drawing on and restructuring Arendt's analysis, retains the term 'totalitarianism' in *The Presidential*

*Papers* (1964; St. Albans: Panther, 1976) but suggests that 'totalitarianism is better understood if it is regarded as a plague rather than examined as a style of ideology' (p. 191). His discussion here precisely elaborates the relationship between Arendt's use of the term and his own, between Nazi totalitarianism and American, and I shall return to it. The phrase 'incipient totalitarianism' will be used in the following discussion. It differentiates the two forms and at the same time indicates the eschatological momentum of contemporary capitalism. As Arendt herself notes at the end of the 1958 edition of *Origins*: 'the crisis of our time and its central experience have brought forth an entirely new form of government which as a potentiality and an ever present danger is only too likely to stay with us from now on' (p. 478).

# Appendix 2

## From Minotaur to Mossmoon

Clive Mossmoon's function as this representative synthesis and the nature of his containment are revealed in the structure of his name. Divided into its three syllabic constituents, it takes us from the erstwhile classic imperial figure of history's most famous Clive, the effective founder of the British Empire in India, to the prospective and exemplary lunar landscape at its incipient totalitarian edge. In the center, absorbed during this neologistic transition, lies the Moss, whose meaning becomes clear in early August 1945 during the Schwarzkommando's synthetic gasoline sniffing sessions at the Leuna petrochemical works' remaining reservoir tanks. There, amongst 'an amazing collection of friends who always seem to show up whenever [Enzian] comes to sniff Leunagasolin,' appears 'the Moss Creature ... stirring like an infant' in response to the fuel's smell. The Moss Creature emerges from the equinoctial, pitching canyons of the Titans – 'the brightest green you can imagine, more burning than fluorescent' – via the mountains of the 'animal north,' the same 'terrible land' of 'the first ancestor' that Blicero's minotaur had retreated into three months before. It embodies the spirit of Pan rising from these invisible depths and is the next term in an aural series – Minotaur: Monitor: Moss Creature: Mossmoon – extending below history's surface to connect the Oven and Rocket State cycles; a 'leaky, wounded' creature shadowed in the 00000 whose pain maintains a subterranean chain of tears linking the death cries of the Minotaur in Pointsman's deepest dreams with the birth cries of the rising self ('a petty indulgent animal') in Clive Mossmoon's 'mired darkness.' As it emerges from the depths, the Moss Creature becomes 'the *peak* that we are allowed to see, the break up through the surface, out of the other silent world,

275

violently,' which fixes itself historically within Clive Mossmoon; and Mossmoon becomes in turn the host of this leaping creature and the container of its struggle inside the spatial, temporal, and structural boundaries of the Rocket State (*GR* 98, 142, 486, 523, 616, 699, 720, 726).

The appearance of the Moss Creature en route to its historical embodiment via the simultaneous and hardly coincidental destruction of Hiroshima and Nagasaki announces to the Schwarzkommando the first appearance of the new order. As Enzian's assistant Christian prepares to shoot the Moss Creature, they both sense its origins in the breath of their first ancestor Mukuru and in the Omumborombanga or creation tree of the Herero, from which it sprouts, as an 'awful branching: the two possibilities already beginning to fly apart at the speed of thought – a new Zone in any case, now, whether Christian fires or refrains.' When Enzian knocks Christian's barrel aside he ensures that military options are restrained by civilian caution. In consequence, 'both men [see] the new branches' and know that the post-war structure of power is starting to manifest itself: 'The Zone, again, has just changed, and they are already on, into the new one ...' (*GR* 524).

Not coincidentally, at the same time in London (early August 1945), Clive Mossmoon 'feels himself rising, as from a *bog* ... delivered onto the sober shore of the Operation, where all is firm underfoot ...' (emphasis added). He has absorbed the Moss or Bog Creature and in the process has completed the formation of his Rocket State character. This character is presented symbolically as a piece of graffiti (appropriately scratched on a toilet or *bog* wall) by Tyrone Slothrop shortly before his dissolution (*GR* 624). The sign represents, in one sense, the imperial order absorbed into the totalitarian, and in another the unstable tensions between private and public dimensions of power characteristic of the post-war order. An inner circle is held between a central point (where, according to Sir Stephen Dodson–Truck, the child's independent ego begins) and an outer ring (where that ego absorbs the public space as it does in the cases of Pointsman and Blicero), between personal *Führer*-dom and passive extinction. The point at which the inner circle reaches the outer, and hence vanishes, equates to the delta–t at which the descending rocket (which the sign also represents) reaches the top of one's head – the point of extinction so dreaded by Pirate Prentice, Franz Pökler, and others (*GR* 7, 206, 425, 616, 624).

# Index

(Names in italics refer to characters from *Gravity's Rainbow*)